Aircraft Hydraulic
Systems

Tornado ADV (F.2) Carrying four Sky Flash missiles

Aircraft Hydraulic Systems

An Introduction to the Analysis of Systems and Components

W. L. Green
Department of Aeronautical and Mechanical Engineering
University of Salford, UK

A Wiley–Interscience Publication

JOHN WILEY AND SONS

Chichester · New York · Brisbane · Toronto · Singapore

Library of Congress Cataloging-in-Publication Data:

Green, W. L. (William L.)
    Aircraft hydraulic systems.

    "A Wiley-Interscience publication."
    Bibliography: p.
    Includes index.
    1. Airplanes—Hydraulic equipment. I. Title.
TL697.H9G68 1985   629.135   85–12385
ISBN 0 471 90848 7

British Library Cataloguing in Publication Data:

Green, W.L.
    Aircraft hydraulic systems: an introduction
    to the analysis of systems and components.
    1. Airplanes—Hydraulic equipment
    I. Title
    629.134′38   TL697.H9

ISBN 0 471 90848 7

Printed and bound in Great Britain.

Contents

viii

Preface

The design of aircraft hydraulic components and systems presents a stimulating challenge to the creative ability of engineers. It poses problems subject to the most severe constraints imposed by airworthiness requirements, environmental restrictions, and reliability criteria. It is also true that the solution of these problems often leads the way to advances in fluid power engineering in many other fields.

Since the publication of the excellent series of books edited by H. G. Conway in the late 1950s there has not been a text which deals with this specific subject. The author therefore felt that after some 30 years as a designer and teacher in this area he could contribute by presenting a new text written from a somewhat different standpoint which would also augment those points mentioned above.

The book is divided into two parts and it is expected that Part 1, which introduces the reader to the aircraft general services and flying control systems and their common components, ending with the analysis of the flying control unit neglecting compressibility of the fluid, would be suitable for students in the penultimate year of an engineering degree course. It involves no mathematical knowledge beyond a simple differential equation. Indeed it can stand in its own right. Part 2 introduces fluid compressibility and uses the method of 'linear perturbations' to analyse the performance of powered flying controls. It discusses stability in some detail for both mechanically and electrically signalled units. Methods for component selection are presented and a brief introduction to non-linear behaviour is given.

The most important innovation is the incorporation of worked examples, self-assessment exercises and numerical questions, features which the author feels are essential in any teaching textbook, whether used in a lecture room or for self-study. There is no attempt to present complicated circuit diagrams or detailed cross-sections of components as these are only fully understood if one initiates a design or lives with them for a considerable time. The selection of materials is not included as there are many texts dealing with this aspect.

The author therefore does not pretend to provide a complete design guide but does hope that the book will serve to introduce young engineers to a subject which holds a prime position in aircraft design. The references will enable the reader to extend his knowledge to whatever extent he requires and in particular the non-linear techniques which have become easier to incorporate with the normal design use of computers.

In the end it is only practice and the help of experienced colleagues together with a thorough appreciation of the problems and their analysis that will produce the required results, and so the author can only hope that this book will inspire engineers entering this field to move the frontiers ever forward.

W. L. GREEN

Acknowledgements

To Professor T. R. Crossley, the then Chairman of the Department of Aeronautical and Mechanical Engineering, for his help and encouragement and to many colleagues past and present. In particular to Mr K. G. 'Peter' Hancock and Mr D. M. 'Don' Bruce amongst others too numerous to mention of the former Electro Hydraulics Ltd., Warrington, UK, where the author spent some of the most interesting and enjoyable years and was fortunate enough to commence his career as assistant to the two outstanding and patient engineers named above.

Appreciation is due for all the secretarial assistance of Miss B. Begal and to Mrs J. Nicholls for her impeccable tracing skills.

Acknowledgement is made to the following companies, individuals and organizations enabling both diagrams and questions to be reproduced by permission of:

British Aerospace plc: photographs and Figs. 1.2 and 1.3.
Dowty Rotol Ltd.: Fig. 4.6
Dowty Hydraulic Units Ltd.: Fig. 8.3
Rexroth Gmbh: Fig. 4.8 and 5.4
Chapman and Hall: H. G. Conway, *Aircraft Hydraulics Vol 2*, Fig. 14.3.
Pitman Publishing Company: H. G. Conway, *Fluid Pressure Mechanisms*, Fig. 11.2
D. M. Bruce: Figs. 4.4, 6.3, 6.5, 8.1 and 18.2
The University of Salford: questions from examination papers.

List of Symbols

K	–	system gain	
L	–	load on servo jack	N
		or length of lever arm	m
		or length of tube	m
		or preset spring load	N
M	–	mass	Kg
N	–	bulk modulus	N/m^2
		or number of pump cylinders	
P	–	hydrostatic pressure	N/m^2
p	–	incremental pressure	N/m^2
P_L	–	load pressure	N/m^2
P_s	–	system pressure	N/m^2
ΔP	–	pressure difference	N/m^2
q	–	incremental volumetric flow	m^3/s
Q	–	volumetric flow rate	m^3/s
Q_c	–	compressibility flow rate	m^3/s
Q_L	–	leakage flow	m^3/s
r	–	(ω/ω_h)	
R	–	reciprocal of valve resistance	1/Ohm
S	–	jack or pump stroke	m
		or spring rate	N/m
s	–	d/dt	1/s
t	–	time	s
		or cylinder wall thickness	m
T	–	torque	Nm
		or servo time constant (closed)	s
		or temperature	°C
u	–	valve lap	m
		of flow velocity	m/s
V	–	volume	m^3
		or voltage	volts
w	–	fraction of annular circumference	
W	–	external work	joules
x	–	incremental displacement	m
		or valve opening	m
X	–	displacement	m
		or valve opening	m
Y	–	transfer function (e.g. $Y(s)$)	
Y	–	displacement	m
Y_1	–	input displacement	m
y	–	incremental displacement	m
Y_0	–	jack displacement	m
z	–	elevation	m
		or fraction of air in oil	
α	–	angle of swash plate in pump	rad

	—	or coefficient of expansion	$1/^{\circ}\text{C m}^3$
β	—	compressibility ($=1/N$)	m^2/N
θ_1	—	desired output movement	m
θ_0	—	actual output movement	m
τ	—	valve time constant	s
ζ	—	non-dimensional damping coefficient	
λ	—	stiffness	N/m
ϕ	—	phase angle	rad
η	—	efficiency	
ν	—	kinematic viscosity	m^2/s
ψ	—	jack diameter	m
ϱ	—	fluid density	kg/m^3
ω	—	rotational speed	rad/s
μ	—	dynamic viscosity	kg/ms
ω_h	—	jack natural frequency	rad/s
(or ω_n)			

NB (Some of these symbols are used with suffixes which clearly define their meaning in the text.

$\bar{\theta}_0$ and $\bar{\theta}_1$ represent increments of θ_0 and θ_1

Part 1

An Introduction to General Service and Flying
Control Systems, and Components. The
Incompressible Analysis of Powered Flying
Control Units

1
Fluid Power and Flight Systems

1.1 Introduction

Fluid power engineering is concerned with the design and assembly of equipment to control the application of mechanical power at places and times where it is required. The diagramatic way in which we can illustrate this is shown in Fig. 1.1. A raw power input derived from either an electric motor or some form of heat engine is transmitted as mechanical power through a shaft to a pump which generates fluid power. This is produced in the form of flow and pressure. Valves are then used so that the power can be fed into actuators to provide controlled mechanical output in the form of force and velocity, or torque and rotational speed. These valves control the direction of movement, the speed, and the maximum force that can be applied by the actuators. The signal to initiate the action and control this speed can be transmitted mechanically, electrically or pneumatically. Indeed, the combination of hydraulics and low power electric signals to control the very high power transmitted by the fluid constitutes the most versatile minimum weight system at present available whether for use on aircraft, robotics, earth-moving or mining equipment, all of which utilize a high power output. The accuracy of control when coupled to a computer station is exceptional.

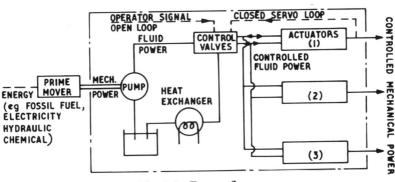

Fig. 1.1 Energy flow

3

To transmit the power we use a fluid which is usually a mineral hydrocarbon or water-based liquid of high chemical stability. The use of pipes results in a flexibility that is absent from mechanical transmissions. As we will also see this system results in a high amplitude force application which is extremely simple in its conception. When compared to any electric element producing a force we are talking in terms of a magnitude advantage greater than 10.

If the system has such advantages of power to weight ratio, flexibility and ease of control, what then are the disadvantages? Like any 'plumbed' system there is a possibility of component failure and leakage. Because we use high pressures the clearances between relatively moving parts are critical and the fluid has to be maintained in a high state of cleanliness. The problem of sealing between relatively moving surfaces such as a piston and cylinder are quite severe. However, the development of thoroughly reliable seals capable of withstanding pressure differentials commonly of the order of 400 bar, together with the ability to produce the necessary surface finish on the metal components, are the reasons why there has been such a large increase in the use of fluid power equipment over the last half century. It is the consistent ability to produce seals and well-machined components which has accentuated and accelerated this advance. The efficiency of the transmission is very high over the medium to high kilowatt (kW) range but when we apply low or fractional kilowatts it can suffer from disadvantages compared to the electrical equipment which is now available. It is true that if one considers air as a fluid we should really discuss the parallel development of pneumatic equipment which, because of its mass production capability and low price, exceeds by many orders of magnitude the quantity of hydraulic equipment used throughout the world. We, however, will concentrate on equipment used to apply power usually many times larger than that available from pneumatic equipment which normally operates from shop airlines around 5–8 bar.

1.2 Flight systems

To illustrate the application of fluid power the use on modern aircraft of high pressure hydraulic units is ideal. It provides the elements of a design philosophy which is usually well in advance of applications in other industries for reasons which will become apparent as we proceed.

An aircraft is a flying machine and as such has to be controlled in pitch, yaw, roll and vertical takeoff. These manoeuvres are initiated by moving control surfaces such as the aileron, elevator, and rudder. The 'swing wing' and 'vectored thrust' are also included in this category. To actuate these mechanisms the pilot, or the auto-pilot, initiates signals which are then translated into mechanical motion at the output. The power is provided by an actuator. Thus, we have a system which has the ability to provide power where and when it is required. This constitutes our number one system for the primary flying controls.

In the early days of aircraft flight the pilot's stick and rudder control were

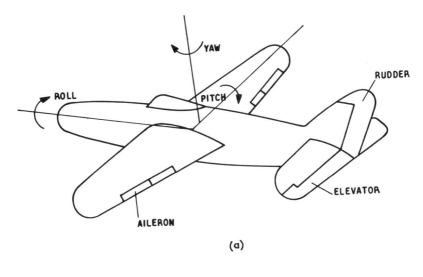

(a)

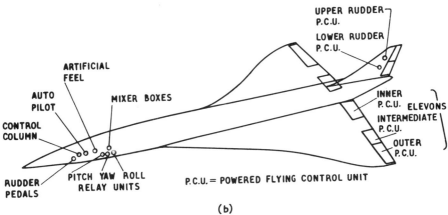

(b)

Fig. 1.2   Flying control surfaces

directly coupled to the surfaces by means of rods, cables, and chains but as the power levels increased it became necessary to provide actuation of the surface against forces well beyond the pilot's capability. This led to the introduction of high pressure hydraulics to transmit the energy. The control of the valves at the actuators was originally mechanical, again, by means of rods and cables taken directly from the cockpit. In modern aircraft the signals are transmitted electrically at low power levels often less than 100 watts (W).

1.3   General or utility services

There is, however, a need for power to be available for other functions of the aircraft which are only required, say, on takeoff and landing or once or twice

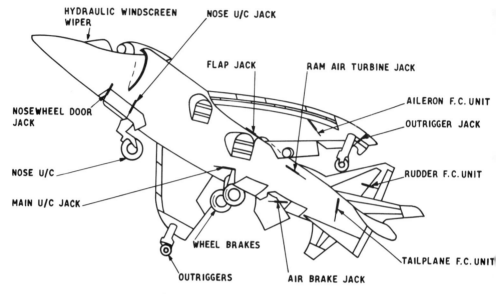

Fig. 1.3 Harrier V/STOL aircraft

in flight. Examples of these are: undercarriage lowering and retraction, wing landing flaps, bomb and cargo doors, air brakes, landing steps, wing fold, refuelling nozzle, nosewheel steering, just to name a few. Here again, high pressure hydraulics provides a solution with the highest power-to-weight ratio and this system is often called the general service or utility system.

Failures in either of these systems cannot be tolerated and it is the extent to which possible failures are covered that in some measure differentiates between the two types. In both of them there must be automatic cover for any failure. In some cases, there is indeed not just duplication but *triplication* of the systems and components. However, the output actuators in the general service system *are not duplicated* and are considered to be in the same class as any other structural member of the aircraft, whereas those in the primary flying control system are indeed duplicated. *This is an extremely important feature to bear in mind.*

It is, of course, system complexity which causes aircraft hydraulic systems and component design to lead the field in this area of technology. Many of the features which were part of aircraft system design in the late 1950s and 1960s are now incorporated in industrial machines but, in the latter, duplication is not required and therefore simplicity rules the day. The less stringent performance requirements result in a significant price reduction, and the material safety factors increase dramatically compared with aircraft components from a value of two to five or more. This leads to increased weight, but in spite of this the advantage still lies with fluid power, *vis-à vis* other systems [1].

1.4 General comments

In order to understand systems fluid power designers have evolved symbols similar to those used in the electrical industry [2]. A brief selection is shown in Fig. 1.4, and the meaning and significance of these will be made clear as we proceed. To follow the fundamental analysis it will be necessary to call upon the subjects of fluid mechanics and control theory together with some solid mechanics, and a basic understanding is assumed.

Although there will be a discussion of the utility or general service system this will act as an introduction to the primary flying control system and its components as a subject of study. In order to do this it will therefore be

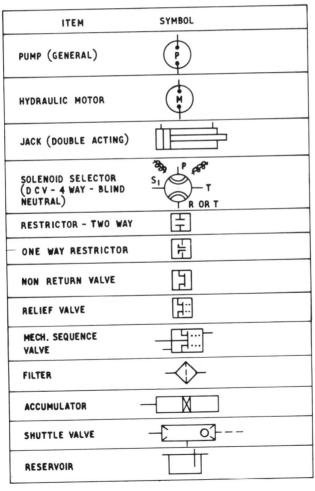

Fig. 1.4 Hydraulic symbols

necessary to cover the miscellaneous components which are part of these systems and to include a moderately detailed treatment.

1.5 Exercises

Self-assessment questions

1. What was the prime factor which resulted in the use of fluid power in aircraft?
2. What modern developments lead to the feasibility of using this equipment?
3. What generally differentiates the two named types of system?
4. Try to think of any alternative sources of energy to provide the primary power to drive a pump if all the normal systems on the aircraft fail.
5. Can you think of any other services in the aircraft in addition to those mentioned above which are suitable for hydraulic actuation?

2
Hydrostatic Systems and Fluid Power

2.1 Pressure and force amplification

The systems that we are now to consider [1] deliver their power mainly through linear actuators described in the aircraft industry as hydraulic jacks, otherwise known as 'cylinders' or 'rams', all based on the original conception of the Bramah press. Compressive stress, that is, hydrostatic pressure, is transmitted through a fluid equally in all directions and hence acts on all internal parts of any vessel such as that shown in Fig. 2.1. In particular, the numerical value of the pressure P is given by the force applied to the piston divided by the area, that is:

$$P = \frac{F}{A} \tag{2.1}$$

If we can produce a pressure P and apply it to a piston free to move axially in the absence of any restraint other than the opposing force, the pressure gives rise to a force and the only limiting factors on the magnitude of this will be the ability of the vessel to withstand the pressure without failure and the prevention of any leakage occurring between piston and cylinder. The simplest demonstration is the single acting jack shown diagrammatically in Fig. 2.2.

The external load F resists extension. The pressure P necessary to overcome this load is then determined by the equation given previously assuming no friction between the piston and the cylinder. A volume of fluid δV is forced into

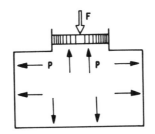

Fig. 2.1   Hydrostatic pressure

9

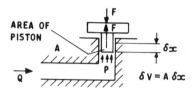

Fig. 2.2   Single-acting jack

the jack and causes it to extend a distance δx in a time δt. The flow rate Q into the jack is

$$Q = \frac{dV}{dt} = A \frac{dx}{dt} \tag{2.2}$$

The fluid performs work $P \cdot \delta V$ on the jack, i.e.

$$\text{work done} = F \,\delta x = P \cdot A \cdot \delta x = P \,\delta V \tag{2.3}$$

$$\text{power} = \frac{P \cdot \delta V}{\delta t} = PQ \tag{2.4}$$

Therefore the fluid supplies power (PQ) to the jack. If the load remains compressive but the jack closes, that is 'retracts', the load F follows the piston and the jack acts as a brake, so the volume δV flows out — that is, fluid power (PQ) is generated; we force fluid out of the jack which acts as a pressure generator or pump.

NB: The pressure P in the jack is determined by the load. The value of Q determines the speed at which the jack moves, and hence if we require a certain velocity the specified flow rate has to be maintained. Figure 2.3 then demonstrates how this ability to transmit pressure results in a force amplification, that is:

$$\frac{F_2}{F_1} = \frac{A_2}{A_1} \tag{2.5}$$

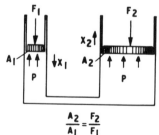

$$\frac{A_2}{A_1} = \frac{F_2}{F_1}$$

i.e. A LEVER EQUIVALENT : $F_2 = \dfrac{F_1 A_2}{A_1}$

Fig. 2.3   Force amplification

2.2 Examples of general principles

If for the jack previously described the load was tensile it would merely pull the piston along the cylinder. How do we cater for this condition? The answer is to use a double-acting jack illustrated in Fig. 2.4 where all the parts are clearly designated.

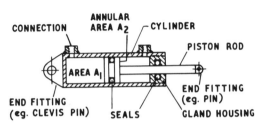

Fig. 2.4   Double-acting jack

The jack cylinder and piston are, of course, cylindrical. The construction of the jack [4] comprises a cylinder manufactured from light alloy or more often high strength steel to a top quality surface finish with a chromium-plated piston rod. Material composition is not the limit on the pressure. The major factor is sealing between the piston and the cylinder or more particularly between the piston rod and the gland housing where relative movement between metallic parts takes place. Indeed, it is the development of the elastomeric seal assembly which has enabled the pressures used in fluid power equipment to be increased to their present high level. Advances in the machining process have augmented this to provide the metallic surface finishes required. Figure 2.5 illustrates some seal combinations in common use and in

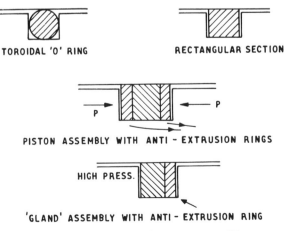

Fig. 2.5   Some sealing arrangements

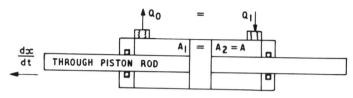

Fig. 2.6 Equal area jack

these the use of backing rings [4–6] prevents the extrusion of the relatively soft elastomeric seal through the clearance between the relative moving parts.

Although the jack illustrated is a double-acting type an alternative, more complicated unit providing equal areas in both directions is most often used in flying controls and is illustrated in Fig. 2.6. It is known as an equal area or tail-rod jack. Naturally, the presence of the seals produces a frictional resistance, normally between 5 and 10% of the load that we would expect from the pressure being applied to the area, and typical values are given in Table 2.1. The equations governing 100% efficient operation are now given by;

$$[\text{Fig. 2.4}] \quad A_1 = \frac{\pi}{4}(\psi_1)^2, \quad A_2 = \frac{\pi}{4}[(\psi_1)^2 - (\psi_2)^2] \tag{2.6}$$

$$\psi_1 = \text{cylinder diameter} \quad \psi_2 = \text{rod diamater}$$

$$P_1 A_1 - P_2 A_2 = \text{force applied}(+\text{compressive}).$$

$$V_1 = A_1 x \quad V_2 = A_2 x \tag{2.7}$$

$$\therefore \quad Q_1 = A_1 \frac{dx}{dt} \quad Q_2 = A_2 \frac{dx}{dt} \tag{2.8}$$

$$\therefore \quad \frac{Q_1}{A_1} = \frac{Q_2}{A_2} \text{ (the continuity equation applied to the jack)} \tag{2.9}$$

$$\text{Work done by the fluid} = P_1 V_1 - P_2 V_2 = Fx \tag{2.10}$$

$$\text{i.e. } P_1 Q_1 - P_2 Q_2 = F \frac{dx}{dt} \tag{2.11}$$

Double-acting jacks with tail-rods have the following properties in addition.

$$A_1 = A_2 \quad \text{and} \quad Q_1 = Q_2 \tag{2.12}$$

TABLE 2.1 Efficiency factors (ηR) relating breakout forces for cylinders of different diameters

Cylinder diameter (mm)	ηR %
20–50	80–85
50–120	85–90
greater than 120	90–95

The strength of the jacks in compression is a critical factor and is covered in references [7] and [8]. They can also incorporate 'internal locks' which are released by the application of fluid pressure when a surface or undercarriage, for example, needs to be locked in position [5]. In addition they can incorporate dampers to slow them down as they approach the limit of their strokes [5], [9].

2.3 Elementary laws of fluid mechanics

Leaving aside hydrostatic pressure and force the two most useful relationships are the continuity and energy equations. Because these have been covered in the basic courses in fluid mechanics [3] they will only briefly be discussed with suitable illustrations being given to help understand their application to our particular case. We will only consider the steady state condition, that is, time invarient or quasi-time invarient systems. Here the continuity equation states that the mass flow rate into the control volume equals that out of it (Fig. 2.7).

$$\sum_1^n \varrho_1 Q_1 = \sum_1^m \varrho_0 Q_0 \tag{2.13}$$

for example
$$\varrho_1 Q_1 = \varrho_2 Q_2 + \varrho_3 Q_3 \tag{2.14}$$

If the density is constant this reduces to

$$Q_1 = Q_2 + Q_3 \tag{2.15}$$

NB: Across the jack it is the velocity of the piston that is constant!
   The energy equation states that between two stations (1) and (2)

$$\frac{P_1}{\varrho_1} + \frac{u^2}{2} + gz_1 + e_1 = \frac{P_2}{\varrho_2} + \frac{u_2^2}{2} + gz_2 + e_2 + W + H \tag{2.16}$$

where
$$e = C_p T \tag{2.17}$$

(C_p—specific heat, T—temperature, z—elevation and u—fluid velocity).

Unless otherwise stated the heat flow H can be ignored. In most of the cases we can consider that the velocity of flow in the pipe line, u, is less than 10 m/s, and any elevation difference is less than 5 m. Therefore the terms involving these variables can be ignored unless we are specifically instructed otherwise. The density ϱ of mineral hydrocarbons is of the order of 870 kg/m^3, and for

Fig. 2.7   Continuity of flow

the sake of the argument let us assume this to be 1000 kg/m^3. The pressure P is in the region of 100–300 bar (1 bar $= 10^5$ N/m^2)

$$\therefore \quad P/\varrho > \frac{100 \times 10^5}{10^3} \text{ i.e. } > 10^4$$

$$\frac{u^2}{2} < 50 \quad \text{and} \quad gz < 50$$

The only significant terms are therefore (P/ϱ) and $C_p T$

i.e. $$\frac{P_1 - P_2}{\varrho} = C_p(T_2 - T_1) \text{ if } W = 0 \qquad (2.18)$$

Because of this predominance of the pressure term these systems are called hydrostatic.

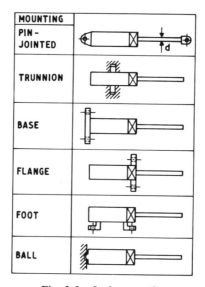

Fig. 2.8   Jack mounting

2.4   Exercises

(A) *Self-assessment*

1. Consider a remote control mechanism to move a ship's rudder consisting of a transmitter (piston in a cylinder) and receiver (an actuator). Can you think of any problem associated with such a mechanism? [5]
2. In late sixteenth century they had trouble when thinking about force amplification. Where did the extra force come from? Do you understand the full explanation?
3. If two jacks are supplied from the same pump and extend against equal loads one always moves first. Why?

4. Heat exchangers are common in aircraft hydraulic systems. Why are they necessary?

(B) *Numerical*
1. A double-acting jack is made to extend. The cylinder diameter is 5 cm and the rod 2 cm. If a pump provides a flow 40 cm^3/s, what is the velocity of piston and the flow rate out of the jack? If the pump flow is applied to the annular side what is the flow out of the full area side and the consequent velocity of retraction? (2 cm/s, 33 cm/s, 47.5 cm^3/s 2.42 cm/s)
2. A compressive load of 10 kN is applied to such a jack. What pressure is required at the full area side to extend the jack assuming the annular area pressure to be negligible and the jack to be 100% efficient? (51 bar)
3. What pressure is required in the full area to resist the retraction of such a jack if it is 90% efficient and a pressure of 100 bar is applied to the annular side? (121 bar)
4. What would be the temperature rise for a pressure fall of 100 bar across a component assuming no heat loss, a density of 875 kg/m^3, and a specific heat of 2×10^3 kJ/kg $\cdot$ °C. (5.7 °C)
5. If the flow in question (4) is 2 litres per second (l/s) what power has to be dissipated by any heat exchanger? (20 kW)
6. A tailrod jack has an annular area of 11 cm^2. Calculate the flow that must be supplied to move it at 20 cm/s and calculate the pressure drop in bar across it if the load on it is 15000 N. (0.22 $\cdot$ 1/s, 136 bar)
7. A double-acting jack is extending against a compressive load of 17000 N. The jack has a full bore area of 16 cm^2 and an annular area of 11 cm^2. A flow of 320 cm^3/s is fed into the jack and there is a valve in the return line to the reservoir giving a pressure of 60 bar at the jack outlet port. Calculate: (a) the linear speed of movement of the jack; (b) the flow from the jack outlet port; (c) the pressure in bar at the jack inlet. (20 cm/s, 220 cm^3/s, 147.5 bar)

Appendix 2.1: Fluid Properties [10]

OM15 (DTD 585) Mineral oil

Specific gravity 0.85
Bulk modulus 18×10^8 N/m^2
Viscosity

	100 °C	5 centistokes
	50 °C	12
	20 °C	22
	0 °C	44
	−40 °C	420

Skydrol 500: Phosphate Ester (lower fire hazard: more corrosive)

Specific gravity 1.15
Bulk modulus 26×10^8 N/m^2

Viscosity	100 °C	4 centistokes
	60 °C	7
	20 °C	20
	0 °C	45
	−40 °C	500

3
Fluid Power Generation

3.1 The power source: hydraulic pumps

The heart of any hydraulic system is the unit which transforms the mechanical input into fluid power and the machine used is a positive displacement pump which provides a flow proportional to the input speed [3]. The pressure is dependent upon the external resistance of the circuit and the load on the actuator. The simplest example is the single piston unit shown diagrammatically in Fig. 3.1.

As the piston moves to the left it draws in fluid to fill the cylinder through the inlet non-return valve. When it moves to the right it forces the fluid trapped in the cylinder through the outlet connection into the system by way of the non-return valve. With a piston cross-sectional area of A and piston stroke of S, during which fluid is delivered, the volume displaced per revolution of the shaft is (AS). If there are N cylinders the volume displaced per revolution is (NAS). It is more usual to work in terms of the volume displaced per radian so that with the input shaft revolving at ω radians per second (rad/s) and a volume displaced per radian of D the volumetric flow rate is

$$Q = D\omega \tag{3.1}$$

(i.e. the volumetric flow rate is proportional to the input speed).

A series of such units disposed radially about an eccentric cam forms the basis for a typical radial piston machine. Alternatively, if the cylinders are disposed

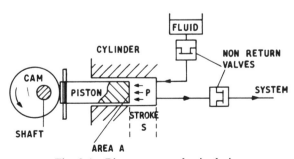

Fig. 3.1 Piston pump—basic design

17

18

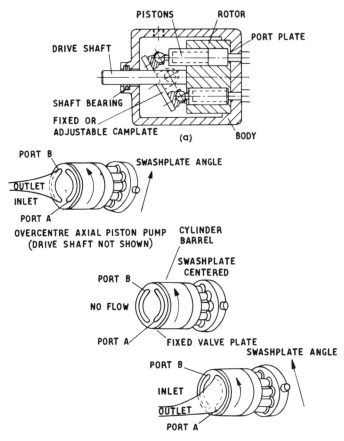

Fig. 3.2   Axial piston pump

as in Fig. 3.2 we have an axial piston or 'swash' type pump which usually does not incorporate any valves in the generally accepted sense but relies upon a slotted valve plate with 'kidney'-shaped ports, as shown, to direct the flow in and out. By changing the angle of this plate the flow rate can be varied from a maximum at one port to zero, when the swash plate is at right angles to the cylinders, to a maximum flow through the other port. Machines can be designed which provide flow to either line and are the basis of most hydrostatic drives. Indeed this is now a standard type of pump used for aircraft applications. It is possible to produce many variations in the detail design. [1]

Other types of pumps are the vane units outlined in Fig. 3.3, which can again be produced as variable delivery pumps, and the simplest type, the gear pump of Fig. 3.4, which incorporates externally meshing gears where the fluid is trapped between the gear teeth and the walls of the housing and is transferred from the inlet to the outlet port. Clearly, increasing pressures, flow rates and reducing weight result in a higher price for the units. Whereas the critical factor

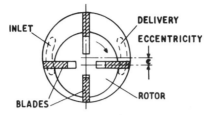

Fig. 3.3   Schematic cross-section of a vane pump

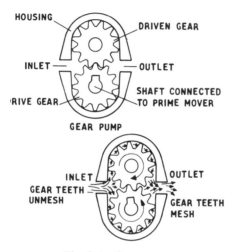

Fig. 3.4   Gear pump

in industrial systems is usually cost, in aircraft systems it is mainly the weight and reliability.

Most pumps used to rotate at between 150 and 300 rad/s but this has been increased in recent years so that standard pumps both of the gear and piston variety can be used up to 400 and 600 rad/s. Characteristics of some typical pumps are then summarized in Appendix 3.2. If the mechanical operation of all these units is analysed it can be seen that they operate on the principle of first offering an increasing volume for the fluid to enter, followed on the delivery stroke by a reducing volume which causes the fluid to be expelled into the system. The simplest general theory was first outlined in Wilson [11] We will follow this treatment.

3.2   Volumetric efficiency

The theoretical output flow of the pump is equal to the product $D\omega$. Within the unit there are leakage paths which are of the laminar flow type (See Appendix 3.1). The flow through these is a function of the pressure drop (Δp)

between the outlet port and the inlet port, the clearance (c), the length (L), width (w), and viscosity (μ) in the form of equation [3]:

$$Q = K \frac{\Delta p . w . \ c^3}{L \ \ \mu}.$$ (3.2)

If all the possible paths are combined then a geometric constant is obtained for the pump, which can be assessed for a particular design, and the leakage flow is given approximately by

$$Q_L = C_s D \frac{\Delta p}{\mu}$$ (3.3)

$C_s D$ represents this geometric constant of the pump taking into account all the leakage losses. The actual flow from the pump can be expressed as:

$$Q_p = Q_T - Q_L$$ (3.4)

i.e.

$$Q_p = D\omega - C_s D \frac{\Delta p}{\mu}$$ (3.5)

If this value is divided by $D\omega$ the volumetric efficiency is defined.

i.e.

$$\eta_v = 1 - C_s \left(\frac{\Delta p}{\mu w} \right)$$ (3.6)

This non-dimensional group ($\Delta p/\mu w$) is extremely important in all pump and fluid motor design. It is the same as the Sommerfeld number for lubricated bearing theory due to the similarity of the problems. A comprehensive discussion is provided in Blackburn [12].

3.3 Mechanical efficiency

The mechanical efficiency is defined in terms of the torque necessary to drive the pump.

$$\text{Power} = T\omega = Q \, \Delta p = D\omega \, \Delta p$$ (3.7)

$$\therefore \quad T = D \, \Delta p$$ (3.8)

The theoretical torque input is thus a product of the cubic displacement per radian and the pressure change across the pump. Of course the actual torque is greater than this value to overcome any losses occurring within the machine. These losses include viscous (T_v), Coulomb friction (T_f), and constant losses (T_c).

i.e. $$T_A = T + T_v + T_f + T_c$$ (3.9)

For most pumps this can be written as

$$T_A = D \, \Delta p + C_D D \, \mu + C_f D \, \Delta p + T_c$$ (3.10)

(T_c can often be neglected). An excellent derivation is given in Merritt [13].

The mechanical efficiency is then

$$\eta_m = \frac{1}{1 + C_f + C_D \dfrac{\mu\omega}{\Delta p}} \qquad (3.11)$$

The overall efficiency is given by

$$\eta_0 = \eta_v \cdot \eta_m \qquad (3.12)$$

Many pumps can be used as hydraulic motors, the input power then being in the form of pressure and flow rate. The analogous equations are

$$\eta_v = \frac{1}{1 + C_s \dfrac{\Delta p}{\mu w}} \qquad (3.13)$$

$$\eta_m = 1 - C_f - C_D \frac{\mu w}{\Delta p} \qquad (3.14)$$

Figure 3.5 is a notional graph of the efficiencies plotted against $(\mu w/\Delta p)$ for a pump and a selection procedure for pumps in general is given in Conway [5]. Most aircraft pumps are now designed to automatically vary the flow from the maximum value to zero once a preset pressure has been attained. This is demonstrated in Fig. 3.6 (see also question 7, p. 24). These provide constant delivery up to the preset pressure, then they provide the delivery required between that pressure and the maximum or 'cut-off' value. They are true variable delivery pumps maintaining a high circuit efficiency.

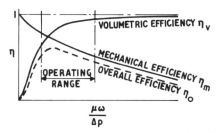

Fig. 3.5   Mechanical, volumetric and overall efficiencies of a hydraulic pump

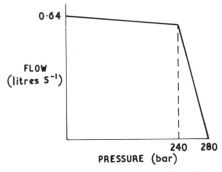

Fig. 3.6   To illustrate question 7

3.4 Example

A hydraulic motor produces 7.5 kW at a speed of 150 rad/s with a pressure loss of 10 bar from pump to motor. Determine the input power to the pump, the pump flow rate, and input torque at 150 rad/s. Efficiences of the pump and motor are given in Table 3.1 below:

TABLE 3.1

Efficiency	Pump	Motor
Volumetric ($p^{\eta}V$)	0.97	0.9
Mechanical ($p^{\eta}M$)	0.95	0.85

System pressure = 90 bar $\times 10^5$ N/m^2
Leakage in system = 6 litres/min = 10^{-4} m^3/s

Solution

The motor torque output = power/speed = $(7.5 \times 10^3)/150 = 50$ Nm. (Static output torque is probably only 90% of this value, say 45 Nm); theoretical torque output = actual torque/mechanical efficiency = $50/0.85 = 59$ Nm.

Pressure available at motor = pressure at pump (that is, the system pressure) − losses = $90 − 10 = 80$ bar = 80×10^5 N/m^2.

Cubic capacity of motor, $Dm = 59/(80 \times 10^5) = 7.4 \times 10^{-6}$ m^3/rad = 7.4 cm^3/rad.

Theoretical flow into the motor = (cubic displacement/rad) $\times$ (rad/s) = $7.4 \times 10^{-6} \times 150 = 1.1 \times 10^{-3}$ m^3/s = 1.1 litres/s.

Actual flow into motor = theoretical flow/volumetric efficiency = $1.1 \times 10^{-3}/0.9 = 1.23 \times 10^{-3}$ m^3/s.

Flow out of pump = actual flow into motor + system leakage = $(1.23 + 0.1) \times 10^{-3} = 1.33 \times 10^{-3}$ m^3/s = 1.33 litres/s.

Fluid power developed by pump = flow $\times$ pressure

$$= (1.33 \times 10^{-3})(90 \times 10^5)$$
$$= 12000 \text{ W} = 12 \text{ kW}$$

This value can be obtained directly from the expression:

(Output power of motor)/$[(m^{\eta}v \times m^{\eta}m \times p^{\eta}v \times p^{\eta}m) \times$ (system efficiency)] = $(7.5 \times 10^3)/[0.9 \times 0.85 \times 0.97 \times 0.95 \times (80/90)] = 12$ kW
(*Note* that the last term in the denominator is the ratio of pressure available at the motor to the system pressure).

Theoretical flow from pump = actual flow/volumetric efficiency

$$= (1.33 \times 10^{-3})/0.97$$
$$= 1.38 \times 10^{-3} \text{ m}^3/\text{s}$$

Therefore, cubic displacement/rad = theoretical flow/speed
$$= (1.38 \times 10^{-3})/150$$
$$= 9.2 \times 10^{-6} \text{ m}^3/\text{rad}$$
$$= 9.2 \text{ cm}^3/\text{rad}.$$

3.5 Exercises

(A) Self-assessment

1. Explain the difference between an axial piston and a radial piston pump.
2. Do all piston type pumps require non-return valves?
3. What effect does a blocked inlet filter have on pump performance?
4. Why is the suction line to the pump larger than the outlet pipe?
5. Describe the relationship between rotating speed and fluid delivery rate of a gear type pump.

(B) Numerical

1. A pump driven at 3000 rpm delivers 0.8 litres/s and the resistance in the circuit is such that the pressure at the delivery is 200 bar.
   Calculate: (a) the power required to drive the pump; (b) the driving torque; (c) the driving torque with the same delivery pressure but with the pump speed dropped to 2000 rpm with a corresponding drop in delivery rate. (16 kW, 50.8 Nm, 50.8)
2. A hydraulic motor performs mechanical work at the rate of 5 kW, calculate the pressure drop across the motor in bar when the flow through it is 1 litre/s. (50 bar)
3. A hydraulic motor is taking a flow of 0.5 litre/s and rotating at 2000 rpm. The meaured pressure drop between its ports is 120 bar. Calculate the torque that it provides. (28.6 Nm)
4. A hydraulic motor with a pressure difference across its ports of 180 bar produces a torque of 500 Nm. Calculate: the flow which must be fed to it to cause it to rotate at 1800 rpm. (52.3 cm^3/s)
5. Describe the differences in the characteristics of the following types of hydraulic power transmission: at constant system pressure:

   (a) variable displacement pump and fixed displacement motor;
   (b) fixed displacement pump and variable displacement motor;
   (c) variable displacement pump and variable displacement motor.

   Illustrate with graphs showing the relationship between output torque and power for each system against (motor speed/constant pump speed).
6. It is desired to construct a fixed pump-variable capacity motor circuit to operate between speeds of 5.5 and 31 rad/s. Two pumps are available of capacity 5.3 and 8.0 cm^3/rad respectively and two motors of maximum

capacity 106.0 and 84 cm^3/rad respectively. The available electric motors all run at 100 rads/s and all efficiencies may be taken as 100% initially. Select a suitable pump and motor combination giving reasons for the selection. How will speeds below 5 rads/s be obtained and how will the motor be prevented from overspeeding? What is the maximum torque that can be obtained from this system, assuming a limiting circuit pressure of 103 bar, and at what speed will this value be obtained? If a constant torque of 50% of this maximum value is to be developed determine the maximum speed and power developed.

((a) 106 cm^3/rad, 5.3 cm^3/rad; (b) max torque 1090 Nm at 5.0 rad/s; (c) 1000 Nm at 5.2 rad/s)

7. An aircraft hydraulic pump has the output characteristics shown in Fig.3.6 when running at one fixed temperature at 330 rad/s. The equations governing its operation are:

Output flow Q (litres/s) $= 0.64 - 0.0005 \cdot p \cdot$ up to 240 bar
Input torque T(Nm) $= 0.9\ pD + 18$
where D is the cubic displacement per radian

Plot the graph of power against pressure and then estimate the maximum power output. Determine D if the volumetric efficiency is 95% and the input torque at maximum power for a mechanical efficiency of 90% (240 bar, 2.04 cm^3, 66 Nm).

8. Determine the maximum overall efficiency of a hydraulic pump in terms of $(\mu\omega/\Delta p)$. The pump runs at 150 rad/s with fluid viscosity of 4×10^{-2} kg/ms at 200 bar. The Coulomb frictional torque and viscous torque are equal. The volumetric efficiency is 84%. ($\simeq 68\%$)

9. Calculate the maximum overall efficiency of a pump where:

$C_f = 0.2$, $C_s = 1 \times 10^{-7}$ and $C_D = 5 \times 10^{-4}$ (73%)

Appendix 3.1: Viscous Force and Flow

See Fig. 3.7 a and b.

(a) Force to move surface $= A\,\mu\,\dfrac{du}{dy} = \mu\,\dfrac{u}{c}\,A$

Torque $\alpha\ \left[\dfrac{A}{c} \times \text{arm}\right] \times \mu \times [w \times \text{radius}]$ i.e. $\alpha(C'_D D)\,\mu w$

where $C'_D D$ is a geometric constant

(b) Flow through an annular clearance. $C \ll L$

$$Q = K_1 \cdot \dfrac{\pi d}{12} \cdot \dfrac{\Delta p}{L} \cdot \dfrac{C^3}{\mu}$$

$K_1 = 1$ for concentric cylinders
$K_1 = 1.5$ for totally eccentric cylinders

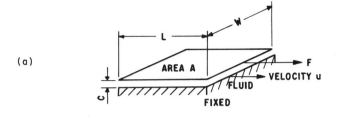

(a)

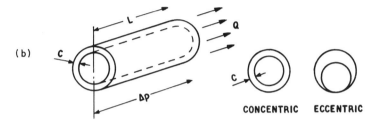

(b)

Fig. 3.7   To illustrate appendix 3.1

Appendix 3.2: Characteristics of Hydraulic Pumps

Characteristics of some typical hydraulic pumps of different designs (F = fixed displacement; V = variable displacement)

Type	Pressure range (bar)	Maximum flow (litre/min)	Speed (rad/s)	Overall efficiency (%)
Gear (F)	20–210	1–700	150–300	60–70
Vane–unbalanced (F and V)	20–65	2–500	120–200	60–70
balanced (F)	20–210	2–1200	200–350	70–80
Piston–radial (F and V) special	20–340	10–800	140–300 (low speed wheel motors)	90–95
axial (F and V)	60–340	2–2000	250–600	90–95
in-line	700	100–5000	3–100	90 +

4

Hydraulic Valves: Control of Direction

4.1 General theory

All valves act as resistances in the line of flow [14, 15]. Within them there are passages and openings through which the fluid has to be forced by the pressure differential across the unit. Most of these resistive paths act in a similar manner to the normal contractions, expansions and orifices [3] in general fluid systems, but in the main their behaviour can be determined by the theory used for orifice flow (Fig. 4.1).

Applying the Bernoulli equation between stations 1 and 3

$$\frac{P_1}{\varrho} + \frac{U_1^2}{2} = \frac{P_3}{\varrho} + \frac{U_1^2}{2} + \frac{\Delta P_L}{\varrho} \tag{4.1}$$

where

$$\frac{\Delta P_L}{\varrho} = C_p(T_3 - T_1) \tag{4.2}$$

and represents the energy loss causing a rise in temperature of the fluid. For a sudden expansion between the minimum area of contraction of the jet a_c and the area A:

$$\frac{\Delta P_L}{\varrho} = \frac{(u_2 - U_1)^2}{2} = \frac{u_2^2}{2} \text{ as } u_2 \gg U_1 \tag{4.3}$$

$$\therefore \quad u_2 = C_v \left(\frac{2}{\varrho} (P_1 - P_3) \right)^{1/2} \tag{4.4}$$

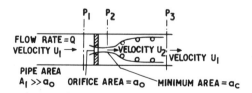

Fig. 4.1   Flow through an orifice

26

(C_v accounts for a minor loss between a_0 and a_c)

$$Q = u_2 a_c = C_c a_0 u_2 \qquad (4.5)$$

$$C_c = \frac{a_c}{a_0} \qquad (4.6)$$

$$\therefore \quad Q = C_q a_0 \left(\frac{2}{\varrho}(P_1 - P_3)\right)^{1/2} \text{ where } C_q = f(\text{Re No}) \qquad (4.7)$$

$C_q = C_c C_v$ and is known as a flow coefficient. It is a function of the Reynolds number (Re No) and over most of the range of flows lies between 0.61 for a sharp-edged entry to between 0.8–0.9 for a rounded inlet. The area open to flow depends on the type of valve element used within the body.

For an orifice, diameter d, Re No $= \dfrac{4Q}{\pi \nu d}$ (ν = kinematic viscosity)

4.2 Valve elements

There are three main type of elements used in aircraft valves [5]: the cylindrical slide (Fig. 4.2a) and a flat side using the 'shear seal' [5] (Fig. 4.2b); the lifting element (Fig. 4.2c) [5], [14], [17], [18]; and the flapper nozzle (Fig. 4.2d) [14], [19].

4.2.1 The cylindrical slide

A cylindrical 'bobbin' or 'spool' comprising several 'lands' connected by a thin stem opens 'ports' to the flow when displaced axially. These ports are in the form of circular holes, a cylindrical annulus or part annular section orifice. For the latter

$$a_0 = w \cdot x \qquad (4.8)$$

where w is the fraction of the circumferential area open to flow. This is then a linear opening valve, that is, the opening is proportional to the displacement x. For a circular port and small openings.

$$a_0 = f(x) \doteqdot K x^{3/2} \qquad (4.9)$$

and is non-linear. There is a large amount of literature on these elements listed in McCloy and Martin [14], and Dransfield [16].

4.2.2 A flat slide using a 'shear seal'

This acts in a similar way to the cylindrical slide except that the sliding portion is flat and the 'shear seal' is forced into contact with it by means of the pressure acting on the annular area of the cylindrical element.

28

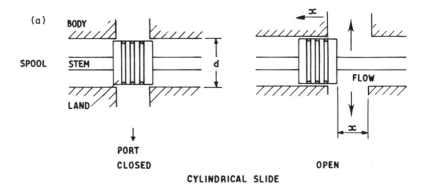

(a)

BODY

SPOOL   STEM   d

LAND

PORT
CLOSED

FLOW

OPEN

CYLINDRICAL SLIDE

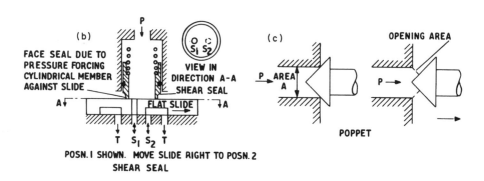

(b)

FACE SEAL DUE TO
PRESSURE FORCING
CYLINDRICAL MEMBER
AGAINST SLIDE

VIEW IN
DIRECTION A-A
SHEAR SEAL

FLAT SLIDE

POSN. I SHOWN. MOVE SLIDE RIGHT TO POSN. 2
SHEAR SEAL

(c)

OPENING AREA

P AREA
A

P →

POPPET

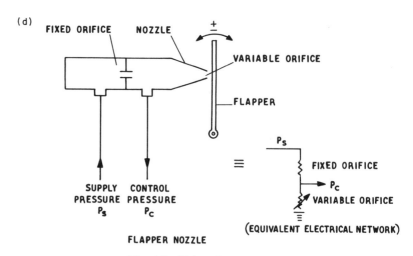

(d)   FIXED ORIFICE   NOZZLE

VARIABLE ORIFICE

FLAPPER

SUPPLY   CONTROL
PRESSURE   PRESSURE
P_s P_c

FLAPPER NOZZLE

P_s

FIXED ORIFICE

→ P_c

VARIABLE ORIFICE

(EQUIVALENT ELECTRICAL NETWORK)

Fig. 4.2   Valve elements

4.2.3 The lifting element

This usually takes the form of a poppet with an included angle between 60° and 90° where the element opens the valve to flow through it by relative movement at right angles to the port.

$$a_0 \doteq \frac{\pi dx}{1.414} \,(\alpha = 90^\circ) \qquad (4.10)$$

for small displacements.

4.2.4 The flapper nozzle

This comprises a 'flapper' which can be moved axially away from, or nearer to, a nozzle as illustrated and is an important component in servo-valve construction which we shall come to later.

4.3 Non-return or 'check' valves

Non-return or check valves (Fig. 4.3) [5] are analogous to electrical diodes. They allow flow in one direction only and comprise a seating element held onto the 'seat' by a light spring. They completely prevent the flow in the opposite direction because of the pressure differential holding the valve closed.

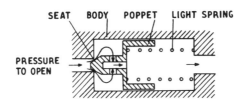

Fig. 4.3   Non-return valve

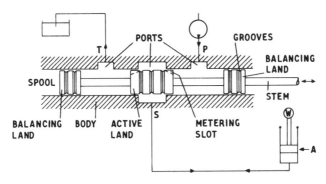

Fig. 4.4   Slide selector three-way, three-position

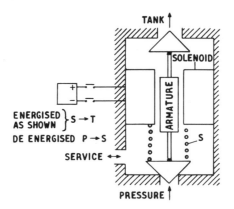

Fig. 4.5   Three-way, two-position poppet valve

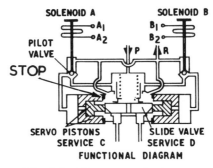

Fig. 4.6   Pilot-operated selector

4.4   Directional control or selector valves

They connect the appropriate fluid lines to ensure the correct direction of actuator movement, and can be signalled directly, either mechanically, or electrically, by means of solenoids [5], [20]. By utilizing a small pilot stage, that is, a small valve which operates on a large spool, the most efficient system of electrohydraulic control can be incorporated. Figure 4.4 outlines the basic elements of a slide selector, while Fig. 4.5 does the same for a solenoid controlled poppet type. Figure 4.6 then indicates the way the slide on a shear seal type (or cylindrical slide) is moved by pistons at the end, this being a mode of operation which utilizes a pilot valve with small solenoids to displace a unit capable of transmitting a high power level. This is the most common type used on modern aircraft systems.

With both solenoids de-energized, the pressure of supply fluid at P holds the pilot valves against the 'return' seats and maintains the balance of the servo-pistons, which hold the slide valve in the neutral position. The two services are then locked hydraulically, as shown in the diagram. The energizing of one

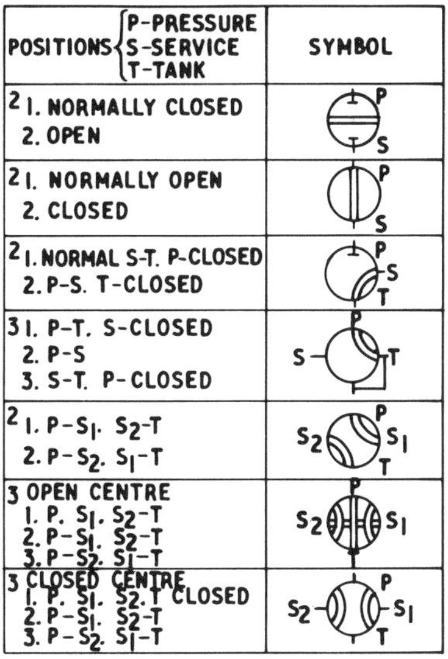

POSITIONS {P-PRESSURE / S-SERVICE / T-TANK}	SYMBOL
2 1. NORMALLY CLOSED 2. OPEN	
2 1. NORMALLY OPEN 2. CLOSED	
2 1. NORMAL S-T. P-CLOSED 2. P-S. T-CLOSED	
3 1. P-T. S-CLOSED 2. P-S 3. S-T. P-CLOSED	
2 1. P-S_1. S_2-T 2. P-S_2. S_1-T	
3 OPEN CENTRE 1. P. S_1. S_2-T 2. P-S_1. S_2-T 3. P-S_2. S_1-T	
3 CLOSED CENTRE 1. P. S_1. S_2. T CLOSED 2. P-S_1. S_2-T 3. P-S_2. S_1-T	

Fig. 4.7 Selector types

solenoid opens the return line R to the servo-cylinder on that side and permits the central servo piston on the opposite side to move the slide valve to one of the operating positions C or D. Fluid flows to the selected service, and at the same time fluid expelled from the remaining service passes to the return line. When the solenoid is de-energized, the pressure in both servo-cylinders is in balance and, due to the arrangement of the stops acting on the outer servo-pistons, the slide valve is moved back to the neutral position.

The major points to be observed when examining a selector are:

1. the number of active connections or ports;
2. the number of positions in which the valve element can be placed;
3. the mode of control, that is, mechanical or electrical.

Figure 4.7 outlines in diagrammatic form some possible configurations. In general, when testing valves to British Standard 4062 the results are plotted in the form of a loss coefficient against the Reynolds Number, but for quick calculation purposes we can assume that the flow between any two ports is of the form

$$Q = K \, \Delta p^{\frac{1}{2}} \tag{4.11}$$

4.5 Note on filtration

It is not possible to give adequate attention to the *extremely important subject of filtration* — cleanliness of the fluid — at this stage [4], [13]. Because valves contain elements with such small clearances between relatively moving

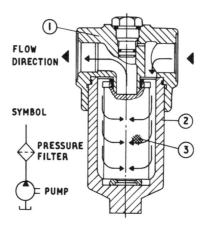

Fig. 4.8   High pressure filter

parts it is necessary to filter out particles of extraneous matter down to 10 microns (μ) (1 μ = 0.001 mm) and 4 μ for servos. Figure 4.8 illustrates a typical modern filter used in the high pressure line. These are extremely complex units capable of withstanding large pressures and a very high number of pulsations in the pressure supply.

The filter shown is suitable for high pressure lines. The filter comprises a filter housing with filter head 1, screw-in filter 2 and the filter element 3. As the pressure filter is exposed to the maximum operating pressure, it must be correspondingly strong and is designed, for example, for a permissible pressure of 315 bar.

4.6   Exercises

(A) Self-assessment

1. What is the purpose of a selector?
2. Show diagrammatically a selector to connect the following passages:

   Position 1 $P-T$, S_1 and S_2 blanked

   Position 2 $P-S_1$, S_2-T

3. Although we have not discussed the forces necessary to displace a cylindrical slide which theoretically should be zero, can you account for the fact that forces do exist and what causes them to arise?
4. Where do you consider the best position to be for a filter?

(B) Numerical

1. Determine the displacement of a spool 12 mm in diameter if the open annular region is half the circumferential length for a flow 1.1 l/s at a pressure difference of 5 bar. Assume $C_q = 0.61$ (2.8 mm)

2. If $C_q = 0.8$ what will the pressure drop fall to? (2.9 bar)
3. At high Reynolds number, the pressure drop across a given orifice restrictor is proportional to the square of the flow rate of fluid passing through it. It is found for this restrictor that experimental results differ significantly from this law only when the flow through the restrictor is at a Reynolds number of 200 or less (based on orifice diameter and the mean flow through the orifice).

(a) Using the assumption that the coefficient of resistance is a function of Reynolds number alone, explain why, with a given pressure drop across the restrictor and providing that the Reynolds number is high, you would not expect the flow through the orifice to depend on fluid viscosity.

(b) The measured pressure drop across the restrictor of 0.5 mm diameter is 50 bar when a flow of 15 cm^3/s of a fluid of 50 cSt kinematic viscosity is passed through it. Use this information to determine:

(i) A point on the pressure drop-flow curve for this restrictor when fluid is used of the same density but of 90 cSt kinematic viscosity, that is for this curve quote a flow and the corresponding pressure drop and show how you have derived these.

(ii) The coefficient of flow at this point if the fluid density is 850 kg/m^3.
(27 cm^3/s, 0.7)

(*Note:* Reynolds number $= (4/\pi)\,(Q/\nu d)$

ν = kinematic viscosity
d = diameter of orifice
Q = flow rate.

1 centistoke $= 1 \times 10^{-6}$ m^2/s)

Buccaneer on take off

5
Hydraulic Valves: Control of Speed and Maximum Pressure

5.1 Flow control valves

Actuator speed is determined by the volumetric flow rate, that is for a jack:

$$\frac{dx}{dt} = \frac{Q}{A} \tag{5.1}$$

When we control the flow rate the actuator moves at the desired speed. To do this two types of valve are used — the plain orifice 'restrictor' and the pressure compensated type. [4], [5], [21] The former behaves in the same way as a simple orifice and the flow characteristics are indicated in Fig. 5.1, that is:

$$Q = K(\Delta p)^{1/2} \tag{5.2}$$

The second type holds the flow rate constant irrespective of the pressure drop within ±5%.

The restrictor can be a simple drilled hole and functions equally well in both directions of jack movement. If it is combined with a parallel check valve it acts unidirectionally as in Figure 5.1b. The flow varies with the pressure drop. As the load often does not change by more than a factor of 2, the pressure difference does not vary by more than 2. It follows that the rate will only

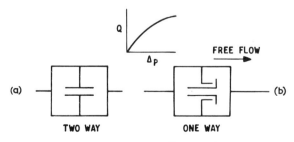

Fig. 5.1   Simple orifice restrictor

change by the ratio of $\sqrt{2}$, in other words by 40%, which is acceptable for most applications. This type of restrictor is a great favourite with aircraft hydraulic system engineers who use orifices down to 1.5 mm with local filtration.

A useful formula for calculation purposes is that,

$$Q = 13a_0 \, (\Delta P)^{\frac{1}{2}} \times 10^{-3}$$
$$\text{(litre/s)} \quad \text{(mm}^2) \quad \text{(bar)}$$
$$\tag{5.3}$$

The use of simple restrictor control is demonstrated in the calculations for the circuit examined in Chapter 6.

5.2 Pressure compensated flow control valves

These combine a fixed and variable orifice which then adjusts itself to provide a constant flow rate as the pressure difference alters (Fig. 5.2) [5], [14], [21]. They maintain an almost constant pressure drop across the fixed orifice producing a constant flow rate. The incorporation of the internal regulating system comprising the variable orifice with pressures acting on the ends of a spool against a biasing spring is the heart of the system. As the valve piston begins to move

$$Q_1 = C_q K (X - x) \left(\frac{2}{\varrho} (P_2 - P_1) \right)^{\frac{1}{2}} \tag{5.4}$$

The flow through the variable orifice must also pass through the fixed orifice so that

$$Q_1 = C_q^1 a_0 \left(\frac{2}{\varrho} (P_s - P_2) \right)^{\frac{1}{2}} \tag{5.5}$$

where C_q and C_q^1 are flow coefficients, P_2 is the pressure between the variable and the fixed orifices and P_L is the load pressure. The force balance on the valve piston results in an equation of equilibrium

$$(P_s - P_2)A = L + Sx \tag{5.6}$$

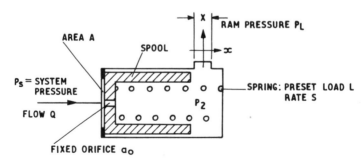

Fig. 5.2   Pressure compensated flow control flow valve

To initiate movement the pressure difference equals the spring load divided by the area. If the spring rate is small then Sx will always be very much less than the preceding term and approximately

$$P_S - P_2 = \frac{L}{A} \qquad (5.7)$$

Substituting this latter value in the previous equation results in the expression

$$Q_1 = C_q^1 a_0 \left(\frac{2}{\varrho} \left(\frac{L}{A} \right) \right)^{\frac{1}{2}} \qquad (5.8)$$

in other words, the flow is almost constant. The position of the piston will adjust itself so that the area open to flow satisfies the equation for changing values of P_L or even P_s. For example, if the maximum variation of

$$L + Sx = 1.1L \qquad (5.9)$$

then the flow variation will be the $(1.1/1)^{\frac{1}{2}} = 1.05$, that is 5%.

Pressure-compensated valves normally give constant flow rates well within this limit and it is possible to combine two such valves to divide a given flow rate in specific proportions.

5.3 Pressure relief valves

These can use the elements indicated schematically in Fig. 5.3. This direct acting type utilizes the force exerted by the pressure on an area to lift a seating element against a preset force applied by a spring. If the load applied by the spring is L, the valve opens when

$$P_c A = L \qquad (5.10)$$

P_c being known as the cracking pressure. Alternatively, a miniature form of this valve can be used to operate a second larger stage in a pilot-operated relief valve [3], [5].

Because of the effects of friction and the chamfer on the valve seat the pressure at which the poppet returns to engage the seat and seal off the flow is within 80–90% of the cracking pressure. In addition the seat chamfer acts in a destabilizing manner [22], [23]. The pressure lifting the poppet off the seat when the fluid begins to creep through the small clearance between the elements causes an opening force in excess of the thrust applied by the opposing spring. The valve then tends to jump open with a subsequent fall in pressure so that the poppet rapidly reseats giving rise to chatter and eventual damage unless adequate damping is provided. Several modifications have been suggested to produce a stable valve with flat pressure flow characteristics [24]. When the valve is open the fluid flows through an orifice similar in shape to the part surface of a cone, or through an annulus when using a cylindrical slide. Relief valves are particularly prone to instability, and it is necessary to select them with great care and particularly to examine their positions in the

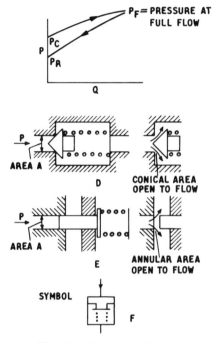

Fig. 5.3   Pressure relief valve

hydraulic circuit. It must be remembered that all the fluid flowing through the relief valve does so with a large pressure fall and rise in fluid temperature. Because of the large pressure drop across the valve the fluid usually 'cavitates', that is 'boils' on the downstream side, with resultant frothing if the fluid is discharged directly into a reservoir.

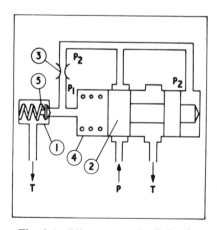

Fig. 5.4   Pilot-operated relief valve

The pilot operated valve illustrated in Fig. 5.4 functions as follows:

Pressurized fluid enters at P and acts on both ends of spool (2). Biasing spring (4) holds it in position and $p_1 = p_2 = P$. When the poppet (5) opens as the pressure reaches the 'cracking' value for the small pilot at p_1 flow occurs through it and $p_2 > p_1$. The main stage (2) then moves under the influence of the pressure difference $(p_2 - p_1)$ [i.e. $(p - p_1)$] and opens the inlet to tank connection i.e. $P \rightarrow T$ to relieve large flows.

The loss characteristics of a relief valve or the 'loss coefficients' are normally given for the pressure drop across the valve when it is open. This will not include the pressure at which the valve cracks open but merely the increase for the changing flow rate.

5.4 Pressure sequencing or maintaining valves

Relief valves can be used for pressure sequencing as indicated in Fig. 5.5, but when the valve is selected the pressure at which the first jack piston (X) moves must be less than the reseat pressure of the relief valve. The poppet or the piston of the valve must be balanced against 'tank' or atmospheric pressure for the valve to remain open when the second jack is moving and a special construction is required which leads to a 'pressure sequencing' or 'maintaining' valve [4], [5]. To give an example: if the valve opens at a pressure of some 140 bar then it must be assumed that the valve will not reseat until the pressure has fallen to 126 bar (X moves; the valve opens then Y moves).

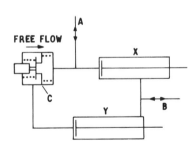

Fig. 5.5   Use of a pressure maintaining valve

5.5 Pressure reducing valves

Sometimes a circuit requires large pressures to operate most of the units but one particular jack only calls for a very low pressure so that small actuators would be required. It is usual to incorporate a pressure reducing valve (Fig. 5.6) [5], [13] so that the pressure applied to this circuit is only a proportion of the maximum pressure. As soon as the pressure rises to a predetermined value equal to the load applied by the spring divided by the area of the piston, in the unit illustrated, the latter will start to move to the right to cut off the

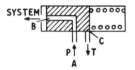

Fig. 5.6   Reducing valve

inlet flow. This 'cut-off' will be effected when the port is completely covered and the pressure in the service line reaches the reduced limit. In most valves any further increase in pressure caused by a sudden 'overload' will cause the piston to move further until it allows fluid to pass from the service line through the port C to tank. When the circuit selector is reset, pressure in the service line must fall below the predetermined reduced value before any movement of the valve piston occurs. There will be the usual hysteresis between these pressures as the valve returns to its original setting [25].

5.6   Mechanical sequence valves

These are similar in construction to non-return valves except that they incorporate a rod which is open to the atmosphere [3], [5]. When this is depressed it moves a poppet off its seat to allow flow in the reverse direction to the normal 'free flow' of the non-return valve element. They are extremely popular units for use in circuits which require the sequencing of one jack after another illustrated in Fig. 5.7 where a door has to be opened before the undercarriage of an aircraft is lowered. Care has to be taken to see that the correct sequencing occurs when the opposite movement for retracting the undercarriage takes place.

5.7   Exercises

(A)   Self-assessement

1. Sketch a valve to maintain an almost constant pressure irrespective of the downstream pressure.
2. Discuss the problem of relief valve 'chatter'.
3. Show by a sketch what is meant by the 'cracking', 'full flow' and 'reseat' pressures of a relief valve.
4. By inserting some values in the equation for the pressure compensated flow control valve see if you can determine how it functions.

(B)   Numerical

1. What spring load is required to produce a cracking pressure of 15 bar with a poppet seat of 12 mm diameter (1.7 kN).
2. If the poppet lifts 2 m what pressure drop is required for a C_q of 0.8 for a flow of 10 litres/s? (233 bar)

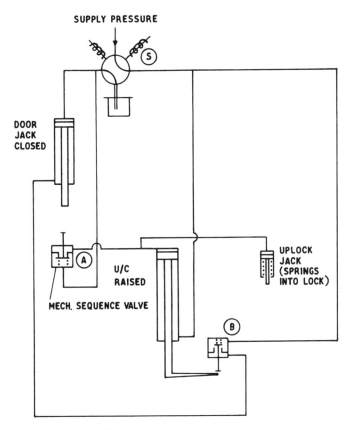

DOOR
JACK
CLOSED

U/C
RAISED

MECH. SEQUENCE VALVE

UPLOCK
JACK
(SPRINGS
INTO LOCK)

TO LOWER: REVERSE SELECTOR S.
DOORS OPEN : A OPENS : U/C UNLOCKS AND LOWERS : B CLOSES.

TO RAISE : SELECTOR AS SHOWN : U/C RAISES: SPRINGS INTO LOCK
: B OPENS : DOORS CLOSE.

Fig. 5.7 Mechanical sequenced undercarriage circuit

3. A jack with equal annular areas on either side of its piston is to be
controlled by a zero lap four-way slide valve which, at a given
displacement, has equal areas of opening at its supply and return ports. The
pressure supply is from a pump with a working pressure range to 280 bar.
The full flow of the pump is 400 cm^3/s, and the maximum value of the
external load on the jack is 60 000 N.

Only the pressure drops at the valve metering ports need be considered
in the circuit calculations and, for each metering port area of 'a' mm^2, a
pressure drop Δp bar across that port results in a flow through it of Q
cm^3/s, where

$$Q = 11\,a(\Delta p)^{\frac{1}{2}}.$$

Return line pressure can be neglected.

(a) Use a pressure of 80% of the working pressure to calculate an annular area for the jack.

(b) Calculate the metering port area at either port required to give a steady speed of 7 cm/s to the jack when it is extending against the stated maximum external load. Calculate the jack speed after the valve has been reversed so as to retract the jack with this maximum load as a following load if, on reversal, the port openings are as previously calculated; explain briefly what happens in the jack. (26.7 cm^2, 3.2 mm^2, 20.9 cm/s)

6
Fluid power circuits

6.1 Constant pressure circuits

Most aircraft pumps incorporate stroke control mechanisms by means of which their displacement D can be varied. This is achieved by altering the angle between cylinder block and swashplate in an axial piston pump, or the eccentricity of the cam ring in a radial one. When such a pump is driven at almost constant speed, via a power take-off from the engines then the stroke control provides a direct control of the flow delivered by the pump [1].

Such a pump directly coupled to a jack or motor as in Fig. 6.1 provides an efficient variable speed hydrostatic transmission. The stroke control sets the flow from the pump, and therefore the speed of the motor, while the load torque controls the pressure drop across the motor and this is almost equal to that across the pump.

Fig. 6.1 Simple pump motor transmission

Unfortunately this circuit requires a separate pump for each jack or motor and is therefore unacceptable for most aircraft applications although it is used in circumstances where pipelines are excessively long.

The usual aircraft circuit is based on a *common supply line*, for all the different services, which is maintained at a constant pressure. This is achieved using a variable displacement pump and applying the pressure at its outlet (the supply line pressure) to a spring-loaded piston which is coupled to the stroke control mechanism: any tendency for the pressure to rise causes the piston to move and reduce the stroke and thereby the delivery of the pump. Such a pump, which is said to be pressure compensated, is the standard one used for aircraft applications, and its characteristics are shown in Fig. 6.2.

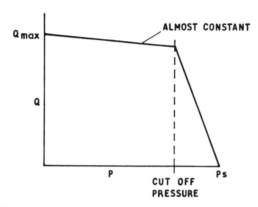

Fig. 6.2   Characteristics of a typical aircraft pump

The action of the pressure compensated pump is demonstrated by the hypothetical circuit, Fig. 6.3, in which the outlet of such a pump is returned to the reservoir via a needle valve. If the valve is closed then the stroke control will move to very nearly the zero stroke position, pressure at A will be maintained at 200 bar and the flow generated by the pump will be only suficient to make up the internal leakage. As the valve is opened to a particular position then the pump stroke will rapidly increase to the point at which the flow through the valve produces a pressure drop across it of 200 bar. As it is opened further then the flow will increase and eventually reach a maximum value corresponding to the maximum pump stroke. Up to this point the pressure at A has been maintained at 200 bar. With the valve opened further still the flow will remain constant and the pressure at A will fall to the value of the pressure drop across the valve at this flow.

We shall generally only consider cases where the flow taken by the circuit is less than the maximum so that the supply line can be considered at a constant set pressure, in this case 200 bar.

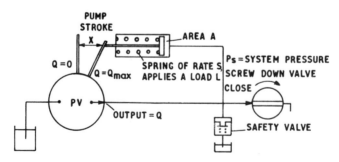

Fig. 6.3   Pump mechanism to provide characteristics of Fig. 6.2

6.2. Speed control and positioning of flow control valves

When the flow from a fixed delivery pump is divided between particular sub-circuits, the problem of correct apportionment arises. It has been found from experience that there are three positions for the flow control valve [14], and these are sometimes combined for certain applications either to reduce the number of valves or to comply with the particular loading specification. 'Meter in' is the first and is illustrated in Fig. 6.4a. The valve is placed in the line to a specific side of the jack to control the 'in-flow'. With a jack that is extended by the applied pressure a valve in this position controls the flow to the piston-head side. It is a useful method if the load opposes the motion (that is, is unidirectional) when extending, shown as F. The stressing pressure is merely the system pressure and a lightweight jack is the outcome. The selector has been omitted for clarity. It will not control the flow if the load acts in the direction of movement as shown by W and the jack will 'run away' with such a 'following' load. 'Meter out' can then be used, and this is shown in Fig. 6.4b. Flow from the actuator is passed through the valve, and in this position loading conditions of all types can be controlled. It must be stressed, however, that the maximum design pressure for the jack must be greater than the system pressure as:

$$P_2 = P_s[A_1/A_2] \tag{6.1}$$

and as A_1 is greater than A_2, this results in 'pressure intensification'. The

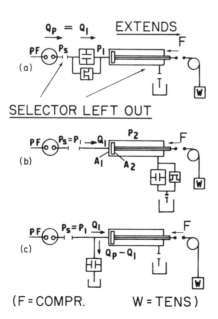

Fig. 6.4   Flow control valve positioning

circuit is useful for loadings of all types, and particularly for infrequent applications at low power.

'Meter by pass' is a third form often used on machine tools, and the arrangement is shown in Fig. 6.4c. Here, jack speed is effectively controlled by returning some of the operating fluid to the tank through a flow control valve. The remaining flow then determines the speed. The pump is connected directly to the actuator. The arrangement will not cater for a following load and is rarely used on aircraft.

When the loading envelope is presented to a designer he must select the flow control valves on the basis of reasoned argument for both directions of movement using a combination of flow control and parallel non-return valves to attain the desired speeds of operation.

6.3 Example

Analyse the circuit of Fig. 6.5 neglecting jack friction, pipe, and selector losses.

If the external load is tensile and therefore resistive $P_A > P_B$

$$P_A - P_B = \frac{1800}{12} = 1500 \text{ N/cm}^2 = 150 \text{ bar}$$

$$\Delta P_1 - \Delta P_2 = 200 - 150 = 50 \text{ bar}$$

Q is the same for each orifice therefore

$$\Delta P_1 = \left(\frac{Q}{13 \times 4}\right)^2 \quad \Delta P_2 = \left(\frac{Q}{13 \times 3}\right)^2$$

$$\frac{\Delta P_2}{\Delta P_1} = \frac{16}{9} \text{ and } \frac{\Delta P_1 + \Delta P_2}{\Delta P_1} = \frac{25}{9}$$

$$\therefore \quad \Delta P_1 = 18 \text{ bar} \quad \Delta P_2 = 32 \text{ bar}$$

$$P_A = 182 \text{ bar} \quad P_B = 32 \text{ bar}$$

$$Q = 13 \times 4 \times (18 \times 10^{-3})^{\frac{1}{2}}$$

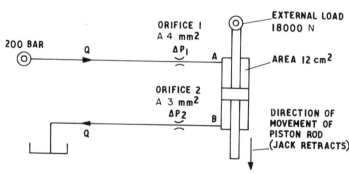

Fig. 6.5 One form of speed control with a constant pressure supply

speed of retraction $\dfrac{0.220.6 \times 10^3}{12} = 18.4$ cm/s

If the external load is compressive and therefore following $P_B > P_A$

$$P_B - P_A = 150 \text{ bar} \quad \Delta P_1 + \Delta P_2 = 200 + 150 = 350 \text{ bar}$$

bar $\qquad \dfrac{\Delta P_1 + \Delta P_2}{\Delta P_1} = \dfrac{25}{9} \quad \therefore \quad \Delta P_1 = 126 \text{ bar } \Delta P_2 = 224$

$$P_A = 74 \text{ bar} \qquad P_B = 224 \text{ bar}$$

$$Q = 13 \times 4(126 \times 10^{-3})^{\frac{1}{2}}$$

$$= 0.5837 \text{ litres/s}$$

speed of retraction $\dfrac{0.5837}{12} \times 10_3 = 48.6$ cm/s

(The questions that you will be asked will normally be simpler than this; either there will be an orifice in only one of the lines or if there is an orifice in both lines these will be of equal area so that $\Delta P_1 = \Delta P_2$).

NB: If the losses in the lines are significant then they have to be calculated and added to the valve losses into and out of the jack. This will be particularly relevant at low temperatures. Appendix 6.1 shows how they can be estimated although many graphs are readily available to ease this task [1], [15]. Appendix 6.2 details acceptable flow velocities in pipes.

6.4 A typical aircraft general service circuit

A simple circuit combining both the power generator and the general services it supplies is shown in Fig. 6.6 [1]. At this stage we have not shown the necessary emergency supply lines to cover failures of the supply pressure or components.

At least two pumps are used drawing fluid from a reservoir. Each pump requires a minimum pressure at the inlet. Flow losses and the loss of atmospheric pressure with altitude necessitate pressure to be applied to the fluid surface in the reservoir. A piston, or engine air, is used to do this. Each pump is then driven by a separate engine and the supplies combined to feed a single high pressure line through non-return valves. This high pressure line feeds all the various subcircuits via non-return valves positioned to prevent flow out of a particular subcircuit when two circuits are operated simultaneously. A further valve which helps relieve any pressure rise, caused by any temperature increase, from trapped fluid between the non-return valve and the circuit selector is incorporated. The selector, for this particular subcircuit, is a three-position 'blind' neutral electrically operated type used to lower and retract the undercarriage, for example. Restrictors are placed close to the jack to control the speed of operation and to cater for the load conditions.

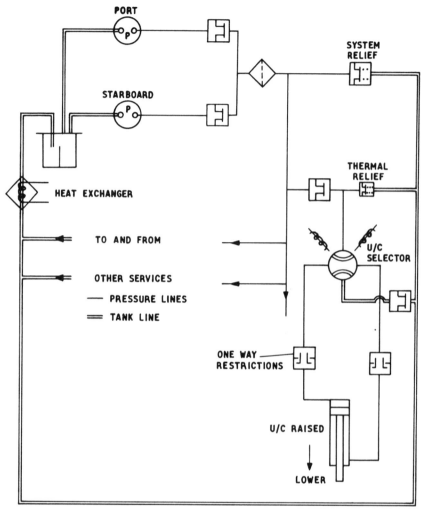

Fig. 6.6 Aircraft general services system (no emergency shown)

A relief valve is also incorporated in the main supply circuit to cover the case of a pump mechanism 'jamming' thus leading to constant flow when there is no place for the fluid to escape. It is set at a value of (1.33 × the system pressure) and acts as a relief to allow the fluid to flow back to tank. Filtration is provided in the high pressure line so that the fluid flowing to the subcircuits is in a clean condition. The fluid returning to the reservoir does so via non-return valves to prevent pressure surges inadvertently actuating valves. The return line is a low pressure line incorporating a heat exchanger usually in the form of a unit transferring heat from the hydraulic oil to the fuel or blown air.

6.5 Exercises

(A) Self-assessment

1. Why is it usual to have the restrictors controlling flow out of a jack?
2. Why do you consider pressurisation of the reservoirs to be necessary? (Section 6.4 para 2).
3. The 'meter bypass' is almost never used on aircraft: why?
4. Where is the point of maximum power input for the pump control described in section 6.1?
5. Why is a heat exchanger necessary?

(B) Numerical

1. Repeat example 6.3 with the orifice 1 ($a = 4$ mm^2) omitted.
     (with tensile load $\Delta P_2 = 50$ bar $P_A = 200$ bar $P_B = 50$ bar
             $Q = 275.8$ cm^3/s speed 23.0 cm/s
   with compressive load $\Delta P_2 = 350$ bar $P_A = 200$ bar $P_B = 350$ bar
             $Q = 729.6$ cm^3/s speed 60.8 cm/s)
2. In the circuit of Fig. 6.5 what happens if orifice 2 is omitted and the load is compressive? (High pressure at B, no speed control).
3. Repeat example 6.3 with both orifices of $a = 3$ mm^2
     (with tensile load $\Delta P_1 = \Delta P_2 = 25$ bar $P_A = 175$ bar $P_B = 25$ bar
             $Q = 195$ cm^3/s speed 16.25 cm/s
   with compressive load $\Delta P_1 = \Delta P_2 = 175$ bar $P_A = 25$ bar $P_B = 175$ bar
             $Q = 515.9$ cm^3/s speed 43.0 cm/s.
4. A tailrod jack has a bore diameter 5 cm and a piston rod diameter 3 cm. A compressive load of 11 000 N acts on it and it is fed from a pump rotating at 2400 rpm which gives a flow of 250 cm^3/s. The simple circuit has been modified so that a pressure dropping valve (a counter-balance valve) has been inserted *between the directional control valve and the reservoir* which provides a pressure drop of 110 bar when there is flow passing through it; the return line pressure at the directional control valve is therefore 110 bar.
     Consider two cases: (A) with the jack extending when the load is resistive: (B) with the jack retracting when the load is following, that is, the jack is acting as a brake. Calculate for each case: (a) the pressures at the jack inlet and outlet ports; (b) the linear velocity of the load: (c) the power required to drive the pump shaft; (d) the torque on the pump shaft. (A: 197.5, 110 bar; 19.9 cm/s; 5 kW; 19.8 Nm) (B. 22.5, 110 bar; 19.9 cm/s; 0.56 kW; 2.26 Nm)
5. The hydraulic jack, efficiency 100%, opening an airbrake extends horizontally against a constant force F, displacing a mass M. The speed is controlled by a restrictor valve at the inlet to the jack comprising a sharp-edged circular orifice. Coefficient of flow is 0.8. The valve can be represented as a resistance R and the flow through it is turbulent. Assuming a constant supply pressure P_s at the valve, from the time movement commences, and

the pressure in the jack to be P_1, write down the equation of motion in terms of the area of jack A and the other designated constants above. Also calculate: (a) a suitable cylinder diameter based on a steady state velocity (the annular side pressure is assumed to be zero and the fluid incompressible). (*Hint:* see (c)); (b) the maximum acceleration (*Hint:* velocity is zero); (c) the diameter of the orifice in the restrictor for a velocity of 6 cm/sec, $F = 5$ kN, $M = 200$ kg, $\varrho = 875$ kg/m^3. $p_s = 300$ bar, $P_1 = 250$ bar for the steady state; (d) The set of conditions above is at $20\,°$C. If the system is tested at $-40\,°$C the time to extend is considerably lengthened. Why is this? All losses except those in the restrictor valve can be neglected. (1.6 cm; 5 m/s^2; 0.42 mm; C_q changes with temperature).

6. The refuelling probe of an aircraft is positioned by a hydraulic jack which extends through a stroke of 0.25 m in 3 s. The speed of the jack is determined by a restrictor which controls the flow into the full-area side of the jack. The restrictor is in the form of a circular orifice which produces a flow rate, Q, given by the expression

$$Q = C_q a \left(\frac{2\,\Delta p}{\varrho}\right)^{1/2}$$

where a is the area of the orifice, Δp is the pressure drop across the orifice, ϱ is the fluid density and the coefficient of flow. C_q is 0.8. The effective aerodynamic load, L, opposing the extension of the probe is given by the expression

$$L = (2 + 64x) \text{ kN}$$

where the displacement of the jack from the fully retracted position is x metres.

Determine a suitable jack-cylinder diameter if the pressure in the jack is 250 bar and the jack efficiency is 90%: it may be assumed that the pressure on the annular side of the jack can be neglected. In addition, calculate the diameter of the restrictor orifice when the pressure at the inlet to the valve is 280 bar and the fluid density is 875 kg/m^3. (*Hint:* an integration is involved). (3.2 cm, 0.7 mm)

Appendix 6.1: Calculation of pressure drops in smooth hydraulic pipe

Density of hydraulic fluid = 875 kg/m^3
   Mean fluid velocity $(u) = (10^3 \times Q)/\pi d^2/4)$
      where Q = flow rate (litres/s)

Pipe diameter (d) (mm)	Fluid viscosity (v) (centistokes CS)	Mean fluid velocity (u) (m/s)
	Calculate Reynolds number (Re) $= 10^3 (ud/v)$	

(1) If Re is less than 2000, flow is laminar, and the pressure drop per unit length is given by:

$$\Delta p/L = 2.8 \times 10^2 [u^2/(\text{Re}.d)]$$

(2) If Re is greater than 2000, flow is transitional or turbulent. Then, find $(\text{Re})^{1/4}$, and the pressure drop per unit length is given by:

$$\Delta p/L = 1.4 \ [u^2/(d.(\text{Re})^{1/4})]$$

Examples

(1) $d = 50$ mm; $u = 1.5$ m/s; $v = 50$ CS
$\text{Re} = 10^3(1.5 \times 50/50) = 1500$
Since Re is less than 2000, flow is laminar,
Therefore, $\Delta p/L = 2.8 \times 10^2 \times (1.5^2/1500 \times 50)$
$\qquad\qquad = 0.0084$ bar/m.

(2) $d = 20$ mm; $u = 10$ m/s; $v = 25$ CS
$\text{Re} = [(10 \times 20)/25] \times 10^3$
$\qquad = 8000$
Since Re is greater than 2000, flow is turbulent.
$(\text{Re})^{1/4} = (8000)^{1/4} = 9.4$

Therefore, $\Delta p/L = 1.4 \times 10^2/20 \times 9.4$
$\qquad\qquad\qquad = 0.75$ bar/m.

Appendix 6.2: Acceptable flow velocities in pipes (m/s)

High pressure: 5–10
Low pressure return: 1–3
Suction lines: < 1

Appendix 6.3: Reservoir pressurization

There is a minimum pressure specified by a manufacturer at a pump inlet p_i (abs)

$$\frac{p_i}{\varrho} = \frac{p_r}{\varrho} + gz - \frac{u^2}{2} - \Sigma \text{ losses between reservoir surface and pump.}$$

and pump.
where p_r = pressure at surface of fluid in reservoir
$\qquad u$ = velocity in pipe
$\qquad z$ = height of surface above pump inlet (negative if below)
$\qquad \varrho$ = liquid density

7
Emergency Operation and Accumulator Performance

7.1 Introduction

As explained previously the purpose of the emergency supply is to ensure that any single failure which can occur does not result in loss of control of the aircraft in flight or when carrying out a landing. In order to demonstrate the way in which this can be done we will examine a simple example of each type of circuit separately.

7.2 General service circuit — emergency operation

Many different circuits are used to provide emergency general service operation using a separate high pressure line [1], [5], [26]. This can be supplied from a simple hand pump, either or both of the two pumps on a twin-engined aircraft, a separate pump driven by an independent power source or from a gas-charged bottle providing pneumatic power. Alternatively a hydro-pneumatic accumulator is used and an analysis of its performance is included in this chapter. In some aircraft pressure supplies are triplicated [29].

As we said earlier the hydraulic jack in this type of circuit is treated as a normal structural member equal to any other part of the aircraft, in other words, a failure of the jack itself is not covered. To overcome the failure of the power source in the general service system we usually provide an emergency source of fluid power via a second set of pipes to 'shuttle valves' at the actuators, illustrated in Fig. 7.1. These valves allow either the normal

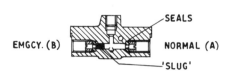

Fig. 7.1   Shuttle valve

52

or emergency supply of fluid to be applied to the actuator and in most cases the speed-controlling restrictors are built into the unit between the shuttle valve and the jack itself so that these do not have be duplicated. The shuttle valve, the restrictor and the jack are built as a single structural member.

At the shuttle valve the power is normally applied to connection A, the other side being vented to the tank pressure. In the event of a failure at A, that is, loss of pressure, then side B is pressurized from the emergency supply. The simplest form of emergency operation is to use a hand pump and this method is still frequently applied on small and medium-sized aircraft.

All necessary valves other than the shuttle and the restrictor are duplicated in the secondary circuit and an example is then shown in Fig. 7.2. The points

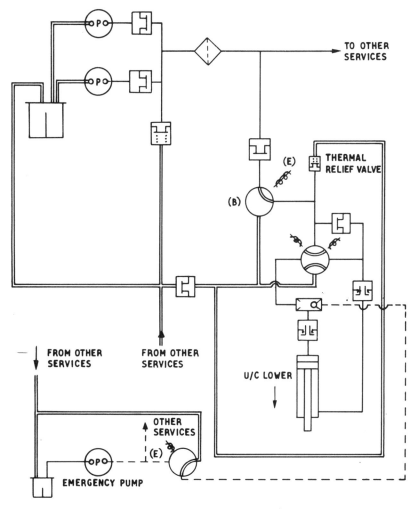

Fig. 7.2 Emergency circuit with standby pump

to notice are:

1. Twin pumps cover the case of a single engine or pump fialure and the non-return valves prevent a flow back from the pump, that is, delivery fluid being forced into the pump that is not working.

2. The release valve, labelled *B*, enables fluid to be forced out of the return side of the jack when emergency operation is required and the normal selector has been jammed in its neutral position. It also covers the case when a normal valve has been jammed in the position for retraction.

3. The emergency selector has its own electrical supply (*E*) and allows pressure to be applied to the jack to move it via the shuttle valve.

4. The emergency pump is usually operated by an alternative energy supply, that is, an air turbine, batteries, etc.

5. The emergency reservoir must contain enough fluid to oeprate all the emergency services before it is drained of fluid.

7.3 Flying control circuit — emergency operation

In this case everything is duplicated (Fig. 7.3). The jacks are designed in series pairs. If a servo-valve is 'jammed' provision must be made to release the fluid out of the side of the jack where failure has occurred. Normally, both jacks will be operating to provide the force to move the actuator but each one on its own must be capable of moving the control surface. This means the circuits are completely disconnected as far as the two servo-units are concerned. This may be confusing in some cases because it looks as though the piston rod in both the jacks is common to each unit. However, a close examination will show that they can be disconnected when necessary. It is, therefore, quite different in concept as far as emergency operation is concerned compared with the general service circuit. Everything is duplicated.

In addition, in some aircraft extra emergency power can be supplied by connecting the general service into the flying control circuit to cover a second possible failure. Very intricate interconnections are incorporated involving power 'take-off' units, in other words, motor pump units. On vertical take-off aircraft there are usually three circuits available and three separate electrical power supplies to cover failure as the aircraft takes off vertically or hovers.

There are a multitude of different types of circuits and some are described in detail in references [27–29]

7.4 Accumulators

Accumulators [5] are the 'capacitance' in hydraulic systems, and can be used to store energy or even absorb energy to eliminate 'surge' pressure or pump-pressure 'ripple'. The action of an accumulator is dependent on the compression and subsequent expansion of a specific mass of gas, held in a cylinder behind a piston (or in industrial applications an elastomeric 'bag') of

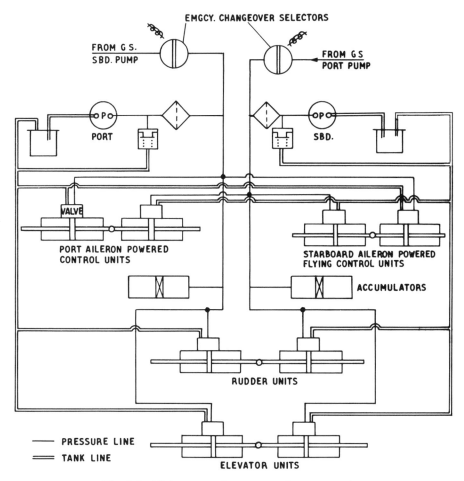

Fig. 7.3   Flying control emergency interconnection

uncharged volume V_A (Figs 7.4 and 7.5). The gas is precharged to a pressure P_c at temperature T_A. In a circuit, this gas is compressed by the flow from the pump to a volume V_s at the maximum system pressure P_s, temperature T_S, as shown in Fig. 7.5. Whatever compression 'law' is followed, the temperature of the gas will then fall towards and, perhaps, reach the initial value, T_A. Consequently, when a valve is operated, the accumulator releases energy into the

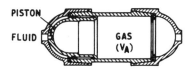

Fig. 7.4   A piston accumulator

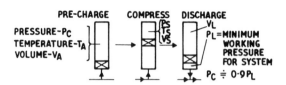

Fig. 7.5 Accumulator operation

system as the gas pressure falls to some minimum value P_L, at temperature T_L, the gas volume then being V_L. The quantity of fluid discharged is:

$$(V_L - V_S) = \Delta V \tag{7.1}$$

The fluid (and hence the gas) is next recompressed to permit the cycle to be repeated.

If the gas employed in such an accumulator was 'ideal' it would be possible to apply the universal gas laws of:

$$PV = RT \tag{7.2}$$

and

$$PV^n = \text{a constant}$$

where $n = \gamma$ for isentropic expansions and is 1.4 for an ideal gas, γ is the ratio of the values of the specific heats and for a real gas is different from this theoretical value of 1.4 [30], [31].

7.5 Example

An accumulator system is required to function in accordance with the cycle indicated in Fig. 7.6. The volume of fluid required by the circuit is given by:

$$(0.5 \times 18) + (2 \times 2) = 13 \text{ litres} = \Delta V = V_L - V_S$$

Therefore, mean flow rate $= 13/20 = 0.65$ litres/s.

If a pump to provide this flow rate is needed, then the accumulator provides the volume of fluid indicated by area X in Fig. 7.6 and it is charged by the pump during the portion of the cycle represented by the area Y. The volume represented by area $X = (2 - 0.65) \times 2 = 1.35 \times 2 = 2.7$ litres; and that corresponding to area $Y = (0.65 - 0.5) \times 18 = 2.7$ litres. Since these 'area'

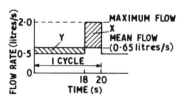

Fig. 7.6 Example

calculations must produce identical results, they serve as a check on the values obtained from the previous steps.

For the purposes of this example, it is assumed that $P_s = 240$ bar and $P_L = 120$ bar. The pressure P_L is the minimum pressure required to operate a service (a jack served by the circuit, for example), allowing for losses between the accumulator and the pressure source. If $P_c = 120 \times 0.9 = 108$ bar (say 100 bar) then assuming isothermal compression, $PV = $ a constant (equation 7.2). Therefore:

$$100 V_A = 240 V_s$$

It is assumed that when fluid is discharged from the accumulator, the gas expands rapidly and consequently adiabatically,

$$PV^{1.4} = \text{a constant}$$

Therefore, $(240/120)^{1/1.4} V_s = V_L$. This expression is combined with the previous equation to obtain $V_s = 4.22$ and $V_A = 4.2$ $(240/100) = 10.2$ litres. Fluid power from the pumps without accumulator $= (2 \times 10^{-3}) \times (240 \times 10^5)$ $= 48$ kW (namely, flow $\times$ maximum pressure). Fluid power from pumps with accumulator $= (0.65 \times 10^{-3}) \times (240 \times 10^5) = 15.6$ kW.

In addition to the smaller pump required using an accumulator, account must be taken of the cost and weight of this unit itself to make an accurate comparison on weight-saving.

7.6   Exercises

(A) Self-assessment

1. Before the advent of high-pressure hydraulic systems how do you think undercarriages were lowered?
2. What other sources of emergency power can be used?
3. How can an accumulator be used for emergency power?
4. How do you think an accumulator reduces pump ripple?
5. Think about the problems of ensuring enough emergency fluid.

(B) Numerical

1. The circuit diagram of Fig. 7.7 is proposed for the sequencing of the normal hydraulic circuit which is to be used for the retraction of an aircraft main undercarriage. The circuit is drawn with the undercarriage locked up and selected up. Door jack A (i) extends to open the doors; (ii) has an internal lock in the closed position; (iii) is subject to a load from the doors which may be in either direction when the jack is extended. (An internal lock is one that exists within the jack and is released by the application of pressure [5]). Undercarriage jack C (i) extends to lower the undercarriage; (ii) is subject in nearly all cases to a considerable tensile load.

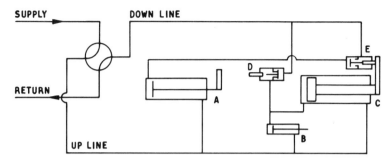

Fig. 7.7 For question 7.1

(a) Explain in detail how the circuit is intended to function on lowering and raising the undercarriage.

(b) Explain the following comments made at a design conference when this proposed circuit was being discussed:
   (i) 'I see that you have used pressure line sequencing on lowering — doesn't this mean that when the undercarriage unlocks on a down selection then the doors may blow back?'
   (ii) 'Haven't we a thermal relief problem at D?'
   (iii) 'I presume that you are looking after retraction on the ground: how?'

(c) Indicate restrictor types and positions for jacks.

(Model answer as follows):

(a) 1. Shown selected up: undercarriage raised, locked; doors closed and locked.
   2. Reverse selector to 'lower': flow to B and C impossible due to sequence valve.
   3. Flow to jack A via valve E; jack extends to open D.
   4. Flow through D to extend B and unlock undercarriage and lower.
   5. Lock jack must be large enough to unlock with pressure on undercarriage jack.
   6. Select 'raise': door jack A cannot move as sequence valve E is closed. B and C retract returning flow through non-return valve integral in D.
   7. B does not need sequencing as it merely releases uplock.
   8. When undercarriage jack C has retracted it opens E and allows door jack to retract and close and lock doors.

(b) (i) No. Flow from A cannot pass through E immediately undercarriage starts to move.
   (ii) Yes. Line from A is blocked by sequence valve. Hence a thermal relief valve must be connected at this point.
   (iii) A weight switch on the undercarriage leg.

(c) Restrictors required in and out of undercarriage jack, that is, two-way each side (restrict flow into door jack).

2. Specify the size of the accumulator required to meet the demand of 0.1 l/s for 12 s and 0.5 l/s for 8 s. The maximum system pressure is 300 bar (abs) the precharge pressure is 180 bar (abs), and the minimim pressure is 200 bar (abs). Use an expansion index of 1.4. (5.5 litres)

3. In a system using a pump with a delivery of 400 ml/s, and a maximum pressure of 70 bar gauge, there is a demand for 0.8 litres of oil over a period of 0.1 s at intermittent intervals. The minimum time interval between demands is 30 s. Determine the size of a suitable accumulator assuming an allowable pressure differential of 10 bar. (10 litres) (Expansion index: use $n = 1.53$)

4. A system is required to deliver 500 ml of oil over a period of 0.1 s with a minimum time between demands of 30 s. Determine the necessary accumulator size if the pump associated with the system has a delivery of 200 ml/s and a maximum delivery pressure of 76 bar (abs). The minimum system pressure is to be 11 bar (abs). Expansion index $n = 1.51$. What capacity pump would be required for this system if no accumulator were to be included? (1.5 litres with $p_1 = 8$ bar; $\simeq 5$ l/s)

8
Servo-Valves

8.1 Throttling directional control valves (servo-valves)

In the general services system the directional selectors have typically three positions, 'extend' — 'inoperative' — 'retract', and the speed is set by quite separate orifice restrictors. If this valve is electrically controlled by two solenoids, as is usual, then with neither solenoid energized it is in 'neutral', in other words, the service is inoperative, while energizing one or the other soldenoid causes it to select the 'extend' or 'retract' position. Both solenoids are never energized together unless there is an electrical fault.

In a flying control system the speed of the service is controlled directly by a mechanical input or indirectly by the magnitude of an electrical current applied to the valve, which is usually a cylindrical slide type (that is, a spool valve). The valve provides both directional selection and the speed control which takes place at the 'metering' ports. Separate orifice restrictors are not required. For reasons which will become apparent later, they are referred to as *servo-valves*. [3], [14]

In Fig. 8.1, x, the displacement of the valve spool from its neutral position, is taken as positive in the direction of the arrows shown. The letter M on the figure denotes the metering orifices for positive values of x when the flow paths are, in (a) $P - S_1$ and $T - S_2$; in (b) $P - S_2$ and $T - S_1$.

If the metering ports are rectangular slots with width, w, as measured around the circumference of the bore, then the orifice area for each metering orifice M is (wx). In the diagrams the metering ports are shown as complete annular grooves machined in the bore so for this case take w as πd.

If with the valve in neutral the metering lands only just cover the metering ports, as in Fig. 8.2, the valve is said to have 'zero lap'. If in neutral they more than cover the metering ports then the valve is said to have 'overlap'. If they do not fully cover the port they have 'underlap'. An overlapped valve requires a finite displacement before flow starts, while an underlapped valve has continuous leakage from pressure to tank when in neutral and the resulting characteristics are shown in the figure.

Typical values of diameter are 5–12 mm, and of the diametral clearance between the spool and the bore 2–8 μm.

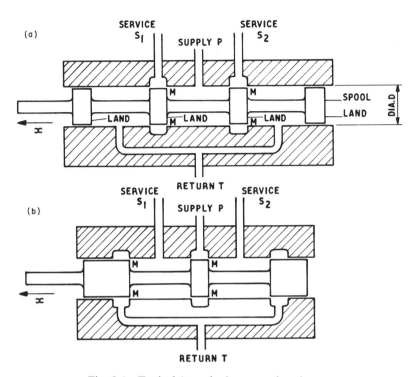

Fig. 8.1   Typical 'zero lap' servo-valve elements

8.2   Electrohydraulic servo-valves

At this stage we will not go into the characteristics of such valves in detail [3] merely stating they produce a flow rate which can be calculated from the equation

$$Q = KI(P_s - P_L)^{1/2} \tag{8.1}$$

where I represents the differential current applied to the torque motor coils of such a valve (see Chapter 16). However, one type of valve is descriptively illustrated in Fig. 8.3 although many variations are described in Shute and Turnbull [32].

The high pressure is first applied at A through the filter element then directly to ports at B, C and to both sides of the nozzle through the fixed orifices at D. The 'flapper' is normally held in the centre position. Each nozzle–orifice combination then applies a control pressure to the end of the main slide E. Moving the flapper changes the value of P_{c1} and P_{c2} in the manner previously illustrated in Fig. 4.2d. If the resistance of the fixed orifice is R_1 and the resistance of the variable orifice is ϵR_1 where $0 < \epsilon \leqslant \infty$ then for a flow Q

$$P_s - P_c = R_1 Q^2 \tag{8.2}$$

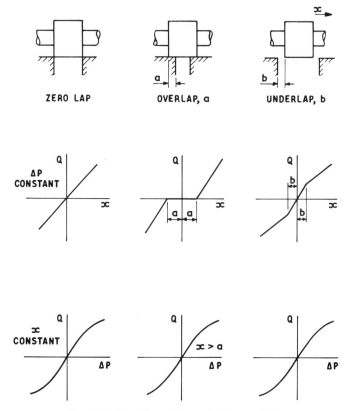

Fig. 8.2   Land/port lap and characteristics

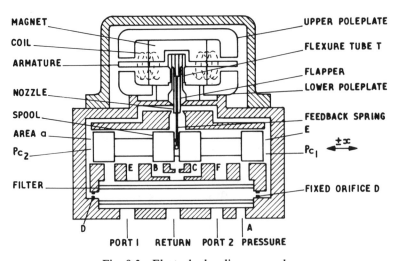

Fig. 8.3   Electrohydraulic servo-valve

$$P_c = \epsilon R_1 Q^2 \tag{8.3}$$

$$\therefore \quad \epsilon = \frac{P_c}{P_s - P_c} \tag{8.4}$$

i.e. $\qquad \epsilon P_s - \epsilon P_c = P_c \tag{8.5}$

$$P_c = \frac{\epsilon P_s}{1 + \epsilon} \tag{8.6}$$

$$(0 < P_c \leqslant P_s)$$

The force equation on the slide is then given by

$$\pm (P_{c1} - P_{c2})A = \text{force} \tag{8.7}$$

The resisting spring force is provided by the flexure tube T and when this balances the pressure force the cylindrical slide will remain in the displaced position. ($\pm x$).

The force is applied to move the flapper by the magnets in the torque motor energized by electrical coils acting on the armature. Here again a force balance is achieved between this force and the pressure force.

The service ports are situated at E and F and the net electrical force is proportional to the differential current. A great deal of work on this type of valve has been carried out over the past three decades and copious references are provided in McCloy and Martin [14].

8.3 Exercises

(A) Self-assessment

1. What is the basic difference between a 'selector' and a servo-valve.
2. Plot $P_c/P_s \sim \epsilon$ for a flapper nozzle. Note any position where approximate linearity occurs.

(B) Numerical

1. A tailrod jack of annular area 11 cm^2 is controlled by a 'zero' lap servo-valve with metering port slots 3 mm wide. The supply pressure is 200 bar and the jack is unloaded. Find the relationship between x, the spool travel in millimetres from neutral and the speed of jack movement in cm/s. Jack friction can be neglected. ($35x$ cm/s) (*Hint:* use Q cm^3/s $= 13a$ (mm^2) (Δp (bar)$^{1/2}$).
2. Repeat 1 for a jack loaded with: (a) a resistive load of 11 000 N; (b) a following load of 11 000 N. ($25x$, $43x$ cm/s)

3. Find the port width w in millimetres required for a zero-lap servo-valve controlling an unloaded tailrod jack with annular area 15 cm^2 from a 210 bar supply if the jack speed in cm/s is to be given by the formula $30x$ where x is the displacement in millimetres of the jack spool from its neutral position. You may neglect jack friction. (3.38 mm)

9
Servo-flying Control Units

9.1 The simple incompressible analysis

These units [3] [33] comprise a servo-valve and actuator coupled together in the manner indicated in Fig. 9.1 which shows a valve with a mechanical input and 'feedback' lever between valve and jack output, or as in Fig. 9.2 where an electrically operated unit is shown incorporating an electrohydraulic servo-valve with electric feedback from potentiometers.

If we consider the valve as a simple set of orifices coupled to an equal area jack as shown in Fig. 9.1 then the following equations can be set down to describe the system considering the fluid to be incompressible, and that no leakage occurs between the supply pressure and the reservoir or across the jack piston. The supply pressure is constant and the return line pressure can be neglected.

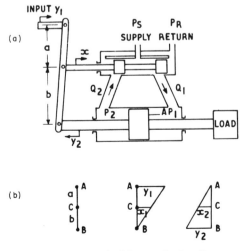

Fig. 9.1 Mechanical lever—hydraulic servo

65

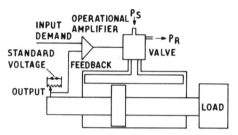

Fig. 9.2   Electrohydraulic servo-system

For a displacement of the valve spool as shown flow into the jack is given by

$$Q_1 = C_q(wx)\left(\frac{2}{\varrho}(P_s - P_1)\right)^{\frac{1}{2}} \tag{9.1}$$

where C_q is the valve coefficient (simple orifice flow) and (wx) the area open to flow for a valve displacement of x. The flow out is given by

$$Q_2 = C_q(wx)\left(\frac{2}{\varrho}(P_2 - P_1)\right)^{\frac{1}{2}} \tag{9.2}$$

As the flows 'in' and 'out' are equal

$$Q_1 = Q_2 = A\frac{dv}{dt} \tag{9.3}$$

where y is the jack piston displacement

$$\therefore \quad P_s - P_1 = P_2 - P_T \doteqdot P_2 \tag{9.4}$$

It is assumed that the return line pressure P_T can be neglected.
For an opposing load L

$$(P_1 - P_2) = \frac{L}{A} = P_L \tag{9.5}$$

(P_L = equivalent load pressure)

$$\therefore \quad P_s - P_2 = P_2 - P_L \tag{9.6}$$

$$\therefore \quad P_2 = \frac{P_s - P_L}{2} \tag{9.7}$$

$$Q_1 = A\frac{dy}{dt} = C_q(wx)\left(\frac{P_s - P_L}{\varrho}\right)^{\frac{1}{2}} = C_q(wx)\left(\frac{Pv}{\varrho}\right)^{\frac{1}{2}} \tag{9.8}$$

$Pv = P_s - P_L$, i.e. *total* valve pressure drop.

$$\therefore \quad \frac{dy}{dt} = \left[\frac{C_q w}{A}\left(\frac{P_v}{\varrho}\right)^{\frac{1}{2}}\right] x = \frac{x}{\tau} \tag{9.9}$$

(*NB*: The time constant τ does really depend upon the load pressure P_L but for many cases can be considered approximately constant). From the equations it can be seen the displacement x results in a jack velocity, i.e.

$$x = \tau \frac{dy}{dt} \tag{9.10}$$

or alternatively,

$$y = \int \frac{x}{\tau} dt \tag{9.11}$$

that is, the output displacement is the integral of the valve displacement, thus the unit is an integrator of time constant τ. For a rotary motor speed ω

$$\tau = \frac{D}{C_q \, w} \left(\frac{P_v}{\varrho} \right)^{1/2} \tag{9.12}$$

9.2 The effect of a 'feedback lever'

If we open a valve and there is no feedback lever, the jack piston will move until it reaches the external or internal limit stops. Let us now consider the situation where we place the feedback lever connected as shown in Fig. 9.1b between the piston rod and the valve input. When lever mechanisms of this nature are involved it is better to consider each pivot fixed in turn than to examine the total effect. Consider B fixed in space and a displacement at A of y_1. This results in an instantaneous valve spool displacement of x_1 where

$$x_1 = \frac{b}{a+b} \cdot y_1 \tag{9.13}$$

This causes a flow into the jack, and the jack piston moves with A fixed, that is, the spool moves in such a way as to 'close off' the flow into the jack as indicated in Fig. 9.1b. The displacement at C is given by

$$x_2 = \frac{-a}{a+b} \cdot y_2 \tag{9.14}$$

and therefore the total instantaneous opening of the valve x is given by

$$x = x_1 + x_2 = \frac{b}{a+b} \cdot y_1 - \frac{a}{a+b} \cdot y_2 \tag{9.15}$$

Also from equation (9.10) x is given by

$$x = \tau \frac{dy_2}{dt} = \left(\frac{b}{a+b} \right) \cdot y_1 - \left(\frac{a}{a+b} \right) \cdot y_2 \tag{9.16}$$

9.3 The general servo-unit equation

We will now use some definitions. We will call y_2 the *actual output* and label it θ_0. We define T as the servo time constant where

$$\tau\left(\frac{a+b}{a}\right) = T \tag{9.17}$$

$y[b/a]$ is the *'desired output'* labelled θ_i. It is really the effect at C of a displacement at A.

$$\therefore \quad \theta_i - \theta_0 = T\frac{d\theta_0}{dt} = E \tag{9.18}$$

$$\text{where } E = \theta_i - \theta_0 \tag{9.19}$$

this error E being the difference between the 'desired' position and the 'actual' position of the output. This equation is a first-order equation and the response to a step, ramp (that is, velocity) and sinusoidal input are shown in Fig. 9.3 with the various characteristics clearly defined.

$\theta_i = \text{step input}$

$$\theta_0 = \theta_i\left[1 - e\frac{-t}{T}\right] = \theta_i \text{ as } t \to \infty \tag{9.20}$$

$\theta_i = Kt$

$$\theta_0 = K\left[t - T\left\{1 - e - \frac{t}{T}\right\}\right] = Kt - KT \text{ as } t \to \infty \tag{9.21}$$

$\theta_i = \theta_i \sin \omega t$

$$\theta_0 = \frac{(\theta_i \sin \omega t - \theta)}{(1 + \omega^2 T^2)^{1/2}} \text{ and } \tan \theta = \omega T \tag{9.22}$$

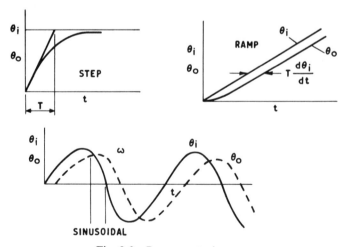

SINUSOIDAL

Fig. 9.3   Response to inputs

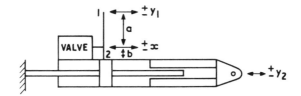

Fig. 9.4   Valve mounted on jack body with lever movements

NB The permissible valve displacement of course is very small. Note that it is only possible to make a step input within this movement of the spool within the body.

Figure 9.4 and question A3 below indicate the more usual lever arrangement for mechanically connected units and Green [34] shows how a poppet valve can be used.

9.4   Why this equation does not provide an adequate model

This equation for the response of the servo flying control unit produces an excellent model but falls down in one respect. It is known that such servo-units can become unstable. A first-order linear differential equation indicates that this is impossible no matter what the input. How then can it occur. This aspect will be covered in the later more advanced analysis of the servo, and as we will see it is due to several causes, the major one being the compressibility of the oil acting as a spring producing a mass–spring–damper 'equivalent' as the mathematical model of the flying control unit.

9.5   Exercises

(A) Self-assessment
1. Why are mechanically connected servo-valves and jacks often called integrators?
2. What happens to the equations if a three-way valve and jack with area ratio of $2:1$ is used?
3. Examine Fig. 9.4. Show that the valve opening is given by

$$\frac{b}{a+b}(y_1 - y_2).$$

(B) Numerical (refer to any 'control' textbook or use reference [3].

1. When the open loop transfer operator is

$$Y = \frac{1}{TD},$$

determine the closed loop transfer function

$$Y_0 \cdot \left(Y_0 = \frac{1}{1 + TD} \right)$$

If T = 0.08 s and the input has a frequency of 1.6 Hz determine the phase shift and amplitude ratio of the output oscillation relative to the input (38.80° lag: 0.779) If on the other hand we only know that the phase lag is 40° at 2 Hz what is T and the amplitude ratio? (0.0668 s, 0.766) If in the last question the amplitude ratio at 2 Hz found on test was 0.70 what conclusion can you draw? (Either the test is wrong or the model is not valid). If with a servo jack T is 0.04 s what is the output positional error when the jack is moving at a constant speed of 11 cm/s? (0.44 cm)

If the output error is 0.6 cm when the jack is moving at 9 cm/s what is T? (0.0667 s)

2. If a jack has an area of 11 cm^2 (that is, take this as the annular area either side of the piston of a jack with a tailrod) with the supply pressure of 210 bar and the return line pressure zero, what is the stalled force of the jack? (You may neglect seal friction.) (23 100 N)

3. The amplitude of a 5 Hz sinusoidal oscillation is 0.3 cm. What is the maximum velocity and acceleration during the oscillation? (9.42 cm/s, 296 cm/s^2)

4. If the jack in question 2, used as a servo-jack, is controlled by a four-way zero lap valve, and if the area of opening of each metering port is 2 mm^2 when the output error is 0.5 cm, then, given that the area of opening is proportional to the linear travel of the valve, what is the time constant of the servo and what is the response to a sudden step displacement of the input of 0.3 cm when the jack is initially stationary and the load is zero?

$$Q = 13a \, (\Delta p)^{\frac{1}{2}} \text{ cm}^3/\text{s}$$
$$\text{when } a \text{ is in mm}^2$$
$$\Delta p \text{ is in bar}$$

Jack area and pressures are as given in question 2:

$$\left[\begin{array}{l} T = 0.0206 \text{ s} \\[2mm] \text{Output } y \\[2mm] = 0.3 \left(1 - e^{-\frac{t}{0.0206}} \right) \end{array} \right.$$

5. The elevator on an aircraft is positioned by an equal area hydraulic jack controlled by a four-way zero lap servo-valve. The actual position θ_0 and desired position θ_i are connected by the relationship

$$\theta_0 - \theta_i = e$$

where e is the valve displacement. If the flow through the valve into and out of the jack is given by

$$Q = 15 \, e(\Delta p)^{\frac{1}{2}} \, 10^{-6} \, \text{m}^3/\text{s}$$

where e is in millimetres and Δp the pressure drop across one land is in bar, show that for an incompressible fluid and ideal jack the system can be described by a first-order differential equation.

Determine the magnitude of the time constant if the constant supply pressure is 300 bar, the load is a constant 12 kN, the jack annular area 6 cm^2. What is the significance of this constant as a measure of the response to a step input and a ramp input of 12 cm/sec. (5.6×10^3 s, $T =$ time to $2/3 \, \theta_i$, 0.67 mm).

Part 2

Fluid Compressibility.   Servo-valve Jack
Analysis and Stability Criteria.
Electrohydraulic Servos.   Choice of Components
and an Introduction to Non-linearity

Third production ADV Tornado AT.00β and ACT Jaguar

10
Liquid Compressibility

10.1 General

In most calculations concerning the application of fluid power the compressibility of the fluid is neglected. Primarily this is because the static pressures involved are relatively low, that is, less than 200 bar. However, some of the most difficult problems in liquid flow, with often disastrous results, arise because of compressibility. It is, for example, the cause of shock waves — known as 'water hammer' or 'surge' — which can produce undesirable oscillations and excessively high pressures in a system.

10.2 Definition of compressibility

The compressibility of a fluid [3], [14] is defined as the change in volume (ΔV) divided by the volume (V) when the pressure is increased by (ΔP)

$$\text{Compressibility } \beta = -\frac{\Delta V}{V} \times \frac{1}{\Delta P} \qquad (10.1)$$

Usually this is expressed in terms of the *bulk modulus* (N)

$$N = -\frac{(\Delta P)}{\dfrac{\Delta V}{V}} = \frac{1}{\beta} \qquad (10.2)$$

The value is not constant and increases with the pressure as shown in Fig. 10.1. Unfortunately, there is an ambiguity here depending on whether the volumetric strain is measured over a finite increase P or an incremental amount Δp, and Fig. 10.2 illustrates this problem. If measured over a finite amount N is defined as the *secant* (S) and if over an increment the *tangent* (T) bulk modulus.

$$\text{i.e. } N_s = -\frac{\dfrac{P}{V_0 - V}}{V} \qquad (10.3)$$

75

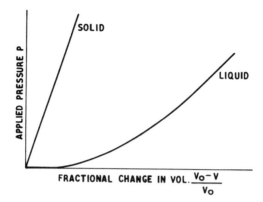

Fig. 10.1   Compressibility

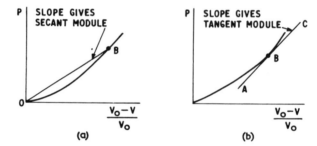

Fig. 10.2   Definition of bulk modulus

$$N_T = \frac{\dfrac{-dP}{dV}}{V} \qquad (10.4)$$

As the liquid is compressed the change can take place adiabatically or iso-thermally and this affects the value of the compressibility. For example, pressure rise zero to working pressure —

isothermal secant (S1) or   isentropic secant (SA)
pressure surge problems — isentropic tangent (TA) if small,
                                            isothermal tangent (TI)

Typically for a mineral oil:

$$N_{S1} = 18.8 \text{ kbar}$$
$$N_{SA} = 21.5 \text{ kbar}$$
$$N_{TA} = 24.1 \text{ kbar}$$
$$N_{TI} = 21.0 \text{ kbar}$$

For a perfect gas, compressed isothermally

$$pV = \text{const} \qquad (10.5)$$

$$\therefore \quad pdV + Vdp = 0$$

$$\therefore \quad N = \frac{-dp}{\frac{dV}{V}} = p \qquad (10.6)$$

If the change is adiabatic (i.e. $pv^\gamma = \text{const.}$)

$$N = \gamma p \qquad (10.7)$$

10.3 Factors affecting compressibility [35]

1. Chemical composition.
2. Temperature.
3. Pressure.

Density $\qquad\qquad \varrho = \varrho_0(1 + ap + bp^2)$

or $\qquad\qquad N = \dfrac{1 + ap + bp^2}{a + 2bp} \qquad (10.8)$

$$a = 4.5 \times 10^{-6}$$

$$b = 5.5 \times 10^{-11}$$

4. Presence of a gas in a liquid [36]. This can occur either as:
    (a) entrained air in the form of bubbles;
    (b) dissolved air in solution.
(A reduction in pressure can cause the gas to escape and is known as 'pseudocavitation'. *True cavitation occurs when the fluid vaporizes at very low absolute pressures. In reality the gas will come out of solution well before the latter event occurs* [37]. Air in solution can be accounted for by the use of an effective bulk modulus, N_e,

$$\frac{1}{N_e} = \frac{1}{N} + \frac{zp_a}{p^2} \qquad (10.9)$$

where z is the fraction of air at atmospheric pressure. Figure 10.3 then demonstrates the effect of air on the value of N_e.

NB (a) This emphasizes that it is essential to remove air from the system, why 'bleeding' is necessary and why 'bleed' points are situated at strategic parts of the system.
    (b) Even with adequate 'bleeding' it has been found that

$$12 \text{ kbar} < N_e < 15 \text{ kbar}$$

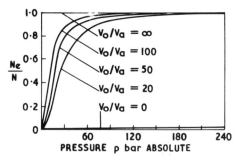

Fig. 10.3 Effect of air on N_e

5. Pipe elasticity leads to the equation

$$\frac{1}{N_e} = \frac{1}{N} = \frac{d}{Et} \qquad (10.10)$$

where

E is the modulus of elasticity of the pipe (or hose)
d is the internal diameter
t is the wall thickness

This hydraulic hoses also add considerably to the system compressibility.

10.4   Some effects of compressibility

10.4.1   Surge pressures [3] [38]

It can be shown that the velocity of a pressure wave in a liquid is given by

$$C = \left(\frac{N_e}{\varrho}\right)^{\frac{1}{2}} \qquad (10.11)$$

and the rise in pressure in a pipe of length L for a sudden valve closure time of

$$t_c < \frac{2L}{C} \qquad (10.12)$$

is given by

$$\Delta p = \varrho u C \qquad (10.13)$$

where u is the flow velocity prior to closure.
$C = 1500$ m/s for oil and if $\varrho = 10^3$ kg/m^3 with $u = 10$ m/s for example, the surge pressure is given by

$$\Delta p = 10^3 \times 10 \times 1.5 \times 10^3 = 150 \text{ bar} \qquad (10.13a)$$

This indicates why, when maximum system pressures were below 100 bar, it was necessary to keep fluid velocities low and only with system pressures above 200 bar can we consider $u \to 10$ m/s. It is essential to keep surge pressures low

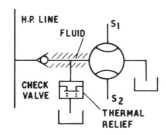

Fig. 10.4   Trapped fluid and thermal relief

even at the expense of longer valve closure times. Surge pressures contribute to the noise problem in hydraulic circuits. The reduction of this phenomena is now the subject of a major research effort [39].

10.4.2   Temperature rise

If the coefficient of expansion is $\alpha = 0.0007/(^\circ C)$, then if a fluid fills a vessel and is prevented from expanding an amount ΔV for a temperature rise of $T^\circ C$.

$$\Delta V = - V_0 \cdot \alpha T$$

$$\therefore \quad \Delta p = \frac{-N \cdot \Delta V}{V} = \frac{N V_0 \alpha T}{V_0} = NT \qquad (10.14)$$

e.g. If $N = 15 \times 10^3$ bar

$$\Delta p = 15 \times 10^3 \times 7 \times 10^{-4}/^\circ C$$

$$= 10.5 \text{ bar}/^\circ C$$

This shows that it is essential to prevent fluid being trapped as in Fig. 10.4, and why in such a case it is necessary to provide a 'thermal relief valve' which is really a small relief valve with good leakproof reseat capability.

10.5   Compressibility flow

When the pressure in a container rises δp in a time δt the volume of fluid decreases by an amount δV

i.e.

$$N = \frac{\dfrac{-\delta p}{\delta V}}{V} \qquad (10.15)$$

or

$$\delta V = -\frac{V}{N} \delta p$$

The rate of volume change as $\delta t \to 0$ can be represented by

$$\frac{dV}{dt} = -\frac{V}{N} \frac{dP}{dt} \qquad (10.16)$$

As this rate of volume change is equal to the flow required to fill the volume in time δt, it is the flow rate necessary to make up this change. This is known as the 'compressibility flow' Q_c [3]

i.e.
$$Q_c = \frac{-dV}{dt} = \frac{V}{N}\frac{dP}{dt} \qquad (10.17)$$

and enables us to express the rate of volume change for this compressibility flow in terms of the rate of pressure rise.

10.6   Exercises

(A) Self-assessment

1. Why does the domestic water supply suffer from 'water hammer' effects when the base pressure is 7 bar?
2. How does an accumulator alleviate surge pressures?
3. Thermal relief valves are necessary in aircraft: discuss why their cumulative leakage is so important.
4. Hydraulic hoses have capacitance. Do they have a significant effect on N?

(B) Numerical

1. Derive equation (10.9)
   If $z = 5\%$, $N = 20$ kbar find N_e at $p = 30$ bar. (9.5×10^8)
2. Find N_e for a steel pipe given $E = 200 \times 10^3$ N/mm^2 and $t/d = 0.1$, $N = 20$ kbar. (18.2×10^8)
3. Derive equation (10.13) by equating the strain energy of the liquid to the kinetic energy of the moving fluid.
4. Calculate t_c for a 20 m long pipe containing oil of specific gravity 0.85 when N_e is 15 kbar $(0.03$ S$)$
5. What volume of fluid would spill through a thermal relief valve for a temperature rise of 50 °C when the trapped volume of oil is 5 cm^3 and N_e is 15 kbar if the valve is set to relieve at 405 bar in a 300 bar system. $(0.014$ cm$^3)$

11
Liquid Springs and Natural Frequencies

11.1 Introduction

The property of compressibility immediately makes feasible the design of a liquid spring. But the problem arises of effectively sealing very large pressure differences as the magnitude of the bulk modulus leads to the conclusion that a very high pressure rise has to be accommodated to produce a consequent volumetric strain and hence an acceptable spring rate. The problem has now been solved by several ingenious sealing methods. However, compressibility also leads to a serious negative effect and that is the natural frequency of a mass bouncing on the column of oil acting as a source of forced resonance.

11.2 The liquid spring

Consider the liquid spring [5, 40] diagrammatically illustrated in Fig. 11.1. Here the piston contains restricted passages across it which allow the flow of oil from one side to the other. As the piston rod is forced inside through a distance x the pressure will rise due to the reduction of space available to the

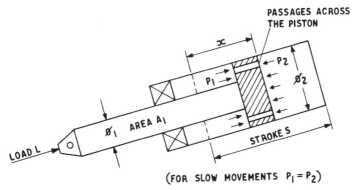

(FOR SLOW MOVEMENTS $P_1 = P_2$)

Fig. 11.1   Liquid spring

81

oil. If the oil volume at any time is V then the definition of the bulk modulus N enables us to write

$$N = - \frac{-dP}{\frac{dV}{V}} \tag{11.1}$$

and $$\int_{P_1}^{P_1} dP = -N \int_{V_1}^{V_2} \frac{dV}{V} \tag{11.2}$$

$$\therefore \quad \frac{P_2 - P_1}{N} = \log_e \frac{V_1}{V_2} \tag{11.3}$$

If P_1 is very small we can state that as $V_2 = V_1 - A \cdot x$ (11.4)

$$\therefore \quad x = \frac{V_1}{A}\left[1 - e^{\frac{-P_2}{N}}\right] \tag{11.5}$$

$$\left(P_2 = N \log_e\left[\frac{V_1}{V_1 - A_1 x}\right]\right) \tag{11.6}$$

Then if we assume a maximum value for P_2 of say 3000 bar with $N = 15$ kbar

$$\frac{A_1 x}{V_1} = 0.182 \tag{11.7}$$

Supposing the displacement x is 0.8 S and

$$V_1 = A_2 S \tag{11.8}$$

$$\frac{\psi_2}{\psi_1} = \left(\frac{A_2}{A_1}\right)^{1/2} \quad \text{i.e. approximately 2 to 1} \tag{11.9}$$

$$\left(NB\text{: If the load is known then } A_1 = \frac{\text{load}}{P_2}\right)$$

This is the basis of the liquid spring used in aircraft undercarriages. Most modern designs utilize a combination of liquid and pneumatic springs to produce the required load deflection curve, in other words the 'oleopneumatic' shock absorber.

11.3   The ultra high pressure seal

One version of this is illustrated in Fig. 11.2 [5]. Here the elastomer is sandwiched between a pressure plate and a steel plate with four projecting pegs. If the area of this plate is A_1 and the pressure p_s then the area of the elastomer in contact with it will be less than this by the area of the pegs, say $A_1 - \Delta A$. Therefore the pressure in the elastomer will be P_R, where

$$P_R = \frac{p_s A_1}{A_1 - \Delta A} \tag{11.10}$$

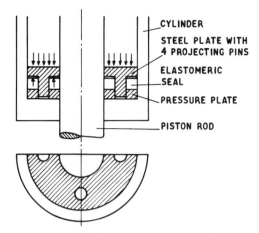

Fig. 11.2   Ultra high pressure seal

It will exceed the fluid system pressure and hence keep the fluid in, although of course there is a large frictional force involved in moving the piston rod. There are now many variations on this type of ultra high pressure seal.

11.4   The natural frequency of a single-acting jack [3]

The change in pressure for a displacement downwards of the piston of $(-\Delta x)$ is Δp where

$$\Delta p = \frac{-N}{V}(\Delta V) \tag{11.11}$$

$$\Delta V = -A\Delta x$$

$$\therefore \quad \Delta p = \frac{+NA}{V}\Delta x \tag{11.2}$$

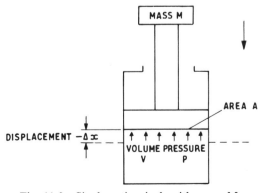

Fig. 11.3   Single-acting jack with mass M

The force equation is given by

$$\Delta p \cdot A = -M \frac{d^2}{dt^2}(\Delta x) \qquad (11.13)$$

$$\therefore \quad \frac{NA^2}{V}\Delta x = -M\frac{d^2(\Delta x)}{dt^2}$$

or

$$\frac{NA^2}{MV}\Delta x = \frac{-d^2(\Delta x)}{dt^2} \qquad (11.14)$$

This is Simple Harmonic Motion with a natural frequency in rad/s of

$$\omega_h = \left(\frac{NA^2}{VM}\right)^{\frac{1}{2}} \qquad (11.15)$$

$$\left(f_h = \frac{\omega_h}{2\pi} \cdot Hz\right)$$

The minimum value occurs when V is a maximum, that is the jack is fully extended. It also indicates the importance of keeping air out of the system in order to maintain a large value of the *hydraulic natural frequency* [ω_h].

11.5 The minimum stiffness of a tailrod jack

For a small movement to the left causing a change of pressure of ΔP_1 and ΔP_2

$$A(\Delta P_1 - \Delta P_2) = -M\frac{d^2(\Delta x)}{dt^2} \qquad (11.16)$$

Also

$$\Delta P_1 = \frac{-N}{V_1}\Delta V_1 \text{ and } \Delta P_2 = \frac{-N}{V_2}\Delta V_2$$

$$\Delta V_1 = -A\,\Delta x \text{ and } \Delta V_2 = A\,\Delta x$$

Substituting in equation (11.16)

$$NA^2\left(\frac{1}{V_1} + \frac{1}{V_2}\right)\cdot \Delta x = -M\frac{d^2(\Delta x)}{dt^2} \qquad (11.17)$$

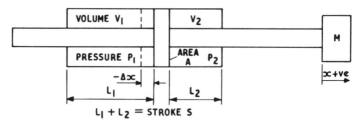

Fig. 11.4 Tailrod: equal-area jack

i.e.
$$\frac{d^2(\Delta x)}{dt^2} = -\omega_h^2(\Delta x)$$

where
$$\omega_h^2 = \frac{NA^2}{M}\left(\frac{1}{V_1} + \frac{1}{V_2}\right) = \frac{\lambda J}{M} \tag{11.18}$$

λ_J is the stiffness of the jack
The minimum value of ω_h will occur when λ_J is a minimum. As $V_1 = AL_1$ and $V_2 = AL_2 = A(S - L_1)$, (S is the stroke)

$$\frac{\lambda_J}{NA} = \frac{1}{L_1} + \frac{1}{S - L_1} \tag{11.19}$$

$$\frac{1}{NA}\frac{d\lambda_J}{dL_1} = 0 \text{ when } -\frac{1}{L_1^2} - \frac{1}{(S-L_1)^2} = 0$$

that is,
$$L_1 = \frac{S}{2} \tag{11.20}$$

Then the minimum value of
$$\lambda_J = \frac{4NA}{S} = \frac{4NA^2}{SA} = \frac{4NA^2}{V_t} \tag{11.21}$$

where V_t is the total volume of the jack (add in the volume between jack and servo-valve where appropriate)

$$\omega_h = \left[\frac{4NA^2}{MV_t}\right]^{1/2} \tag{11.22}$$

Figure 11.5 shows the stiffness λ_J for any position of the piston. If the cylinder has a long exhaust line then the equivalent mass M_e is given by [13]

$$M_e = M + \left(\frac{A}{a}\right)^2 m \tag{11.23}$$

where m is the mass of oil in the line and a is the cross-sectional area of the pipe. For an oil hydraulic motor with *two pipelines*

$$\omega_h = \left[\frac{4ND^2}{V_t J}\right]^{1/2} \tag{11.24}$$

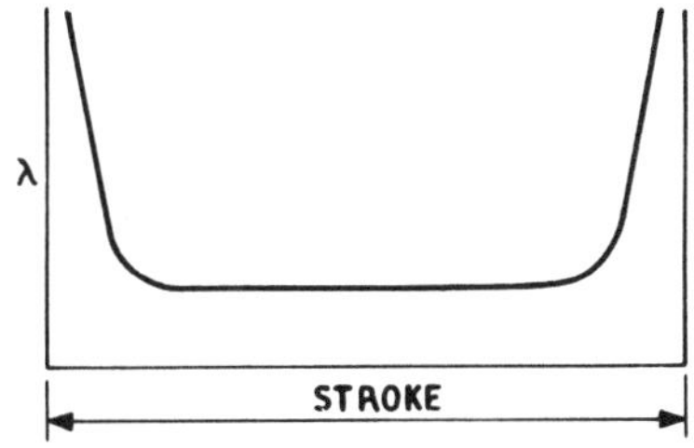

Fig. 11.5 Stiffness of tailrod jack

where D is the cubic displacement per radian and J the inertia of the attached mass.

11.6 Exercises

(A) Self-assessment

1. A liquid spring can be designed to take the weight on the ground. What function will the holes in the piston perform on landing?
2. Can you show the maximum stiffness in Fig. 11.4 to be in the position suggested.
3. Will a hydraulic motor with a single line have the same natural frequency as one with inlet and outlet lines between motor and servo-valve?
4. Why is it important to keep the length of lines between jack and valve short.

(B) Numerical

1. Determine the diameter of the cylinder and piston rod for a liquid spring capable of withstanding a load of 500 kN for a pressure rise of 3 kbar. Calculate the stroke for a cylinder 50 cm long, $N = 15$ kbar. ($\simeq 10$ cm, 4.6 cm, 44 cm)
2. Calculate the value of the minimum natural frequency of a tailrod jack supporting a mass of 2000 kg. The piston diameter is 50 mm and the rod 25 mm. Take $N = 12.4$ kbar and the stroke as 0.45 m. (14.3 Hz)
3. Determine the natural frequency of a single-acting jack supporting a mass of 1052 kg. Effective bulk modulus is 16.5 kbar, the area of the piston is 19.4 cm^2 and the volume trapped 0.6 l. ($\simeq 16$ Hz)

12
Hydraulic Servos: The Compressible Analysis — Part 1

12.1 Introduction

The performance of hydraulic flying control units is analysed using the method of small perturbations, originally applied by Harpur [41] in Great Britain. This takes compressibility into account. A brief resumé of the method is given in Appendix 12.1 and the diagram used for the analysis is that of Fig. 12.1 where a four-way valve is shown controlling an equal area tailrod jack.

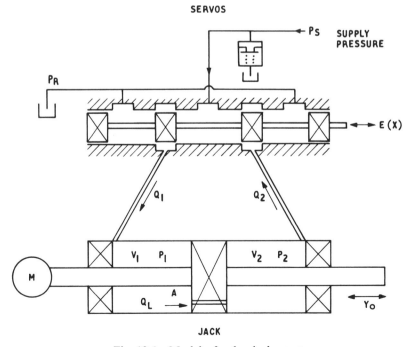

Fig. 12.1 Model of valve jack servo

87

Although this method is adequate as an introduction to the stability criteria and response, it does only deal with small amplitude oscillations and, as the problem is highly non-linear, later models have been suggested to cater for these conditions. In view of the limited nature of this particular text readers are referred to publications which deal with these more advanced aspects in the list of references.

12.2   Assumptions used for the mathematical model

1. The inertia of the fluid is neglected.
2. Fluid density is constant in the orifice equation.
3. Turbulent flow occurs across the valve orifices.
4. The valve openings on both the inlet and outlet side are symmetrical.
5. The only load considered is inertia.
6. The bulk modulus is constant.
7. The perturbations about the mean values are small.
8. The volume included in the final equation takes into account the small passages between jack and valve.
9. The supply pressure is constant and the return line pressure can be neglected.
10. Load disturbances are neglected [14], [42].

12.3   The derivation of the mathematical model [3], [14]

The volumetric flow rates into and out of the jack comprise:

1. That due to piston displacements, in other words, the velocity of the piston multiplied by the area in contact with the fluid (Q_p)
2. The compressibility of the fluid (pressure rises on both sides of piston) (Q_c)
3. Any leakage flow across the jack piston (Q_L)

All the variables are functions of time, for example, $Q_1 = Q_1(t)$ but for the sake of clarity (t) is omitted.

The flow into the jack from the valve port is then described by

$$Q_1 = Q_p + Q_c + Q_L \tag{12.1}$$

$$Q_1 = A\frac{dY_0}{dt} + \frac{V_1}{N}\frac{dP_1}{dt} + Q_L \tag{12.1a}$$

while the flow out is given by

$$Q_2 = Q_p - Q_c + Q_L \tag{12.2}$$

$$Q_2 = A\frac{dY_0}{dt} - \frac{V_2}{N}\frac{dP_2}{dt} + Q_L \tag{12.2a}$$

(P_1 and P_2 are both assumed to rise).

The flow rates Q_1 and Q_2, are governed by the spool valve, that is,

$$Q_1 = \psi(X)(P_s - P_1)^{1/2} \tag{12.3}$$

$$Q_2 = \psi(X)(P_2 - P_R)^{1/2} \tag{12.4}$$

where $\psi(x)$ incorporates the constants 2, ϱ, C_Q and those necessary to change the displacement X into an area.

$$Q_L = \phi(P_1 - P_2) \text{(i.e. a function of the pressure drop)} \tag{12.5}$$

Also we define

$$E = KX = \theta_1 - \theta_0 (K \text{ is a lever constant}) \tag{12.6}$$

E is the difference between θ_1 the 'desired' and θ_0 the 'actual' jack piston position. It is also linearly connected to the valve opening X.

We now express the equations in the form of small perturbations.

$$q_1 = Q_1(t) - Q_1(0) = \left.\frac{\partial Q_1}{\partial X}\right|_i x + \left.\frac{\partial Q_1}{\partial P_1}\right|_i p_1 \tag{12.7}$$

$(\partial Q/(\partial P_s)) = 0$, and all the differentials are taken for $t = 0$; at some mean steady state operating point, say (i))

$$\therefore \quad q_2 = \left.\frac{\partial Q_2}{\partial E}\right|_i x + \left.\frac{\partial Q_2}{\partial P_2}\right|_i p_2 \tag{12.8}$$

and

$$q_L = \left.\frac{\partial Q_L}{\partial P_1}\right|_i p_1 + \left.\frac{\partial Q_L}{\partial P_2}\right|_i p_2 \tag{12.9}$$

12.4 Valve coefficients

From equations (12.3) and (12.4)

$$\frac{\partial Q_1}{\partial X} = \frac{\partial \psi}{\partial X}(P_s - P_1)^{1/2} = C_e = \text{valve flow gain} \tag{12.10}$$

$$\frac{\partial Q}{\partial P_1} = -C_p = \text{valve pressure factor} \tag{12.11}$$

It is assumed that

$$\frac{\partial Q_2}{\partial X} = C_e \tag{12.12}$$

and that

$$\frac{\partial Q}{\partial P_2} = C_p \tag{12.13}$$

and

$$\frac{\partial Q_1}{\partial P_1} = \frac{\partial \phi}{\partial P_1} = -\frac{\partial \phi}{\partial P_2} = \frac{\partial Q_L}{\partial P_2} = C_L = \text{leakage factor} \tag{12.14}$$

In the steady state

$$P_s - P_1 = P_2 - P_R \tag{12.15}$$

12.5 Continuation of the mathematical model

Combining the previous equations and assuming $V_1 = V_2 = V/2$ that is, the piston is at the mid-position — least stiffness (see Chapter 11), and $y_0 = \bar\theta_0$

$$q_1 = C_e x - C_p p_1 = A\frac{d\bar\theta_0}{dt} + \frac{V}{2N}\frac{dp_1}{dt} + C_L p_1 - C_L p_2 \tag{12.16}$$

$$q_2 = C_e x + C_p p_2 = A\frac{d\bar\theta_0}{dt} - \frac{V}{2N}\frac{dp_2}{dt} + C_L p_1 - C_L p_2 \tag{12.17}$$

The equation of motion for the mass is then

$$M\frac{d^2\bar\theta_0}{dt^2} = A(p_1 - p_2) \tag{12.18}$$

(and we can define $p_1 - p_2 = p_L$) \hfill (12.19)

Combining equations (12.1a), (12.2a), (12.13), (12.14), (12.15), (12.16), (12.17) and (12.19) we arrive at the general equation

$$C_e x - \frac{C_p p_L}{2} = A\frac{d\bar\theta_0}{dt} + \frac{V}{4N}\frac{dp_L}{dt} + C_L p_L \tag{12.20}$$

Substituting equation (12.18) results in

$$C_e x - \frac{C_p}{2}\frac{M}{A}\frac{d^2\bar\theta_0}{dt^2} = A\frac{d\bar\theta_0}{dt} + \frac{VM}{4NA}\frac{d^3\bar\theta_0}{dt^3} + \frac{C_L}{A}M\frac{d^2\bar\theta_0}{dt^2}$$

or

$$\frac{C_e x}{A} = \frac{VM}{4NA^2}\frac{d^3\bar\theta_0}{dt^3} + \left[\frac{C_L M}{A^2} + \frac{C_p}{2}\frac{M}{A^2}\right]\frac{d^2\bar\theta_0}{dt^2} + \frac{d\bar\theta_0}{dt} \tag{12.21}$$

In Chapter 11 we showed that the minimum value of ω_h is given by

$$\omega_h^2 = \frac{4NA^2}{VM} \tag{12.22}$$

We let

$$\frac{2\zeta}{\omega_h} = \frac{C_p M}{2A^2} + \frac{C_L M}{A^2} = \frac{M}{A^2}\left[\frac{C_p}{2} + C_L\right] \tag{12.23}$$

Using the previous relationship of equation (12.6), i.e.

$$e = Kx \tag{12.24}$$

$$\frac{C_e x}{A} = \frac{C_e e}{KA} = \frac{e}{\tau} \tag{12.25}$$

where τ = the open loop time constant

ζ = the non-dimensional damping factor

$$\frac{e}{\tau} = \frac{1}{\omega_h^2} \frac{d^3 \bar{\theta}_0}{dt_3} + \frac{2\zeta}{\omega_h} \frac{d^2 \bar{\theta}_0}{dt^2} + \frac{d\bar{\theta}_0}{dt} \tag{12.26}$$

The transfer function can the be written for the open loop as

$$\frac{\bar{\theta}_0(s)}{e(s)} = \frac{\dfrac{1}{\tau}}{s\left(\dfrac{s^2}{\omega_h^2} + \dfrac{2\zeta s}{\omega_h} + 1\right)} \quad \left(s = \frac{d}{dt}\right) \tag{12.27}$$

By definition we have

$$e = \bar{\theta}_i - \bar{\theta}_0 \tag{12.28}$$

i.e. the error between 'desired' and 'actual' output displacement

$$\left\| \frac{\bar{\theta}_i}{\tau} = \frac{1}{\omega_h^2} \frac{d^3 \bar{\theta}_0}{dt^3} + \frac{2\zeta}{\omega_h} \frac{d^2 \bar{\theta}_0}{dt^2} + \frac{d\bar{\theta}_0}{dt} + \frac{\bar{\theta}_0}{\tau} \right\| \tag{12.29}$$

Applying the Routh Hurwitz criterion for stability, that is all the coefficients are positive [3, 14], and

$$1 \cdot \frac{2\zeta}{\omega_h} > \frac{1}{\tau} \cdot \frac{1}{\omega_h^2} \tag{12.30}$$

$$\zeta > \frac{1}{2\omega_h \tau}$$

In terms of the system constants

$$\frac{M}{A^2}\left(\frac{C_p}{2} + C_L\right) > \frac{1}{\dfrac{4NA^2}{VM}} \times \frac{C_e}{KA}$$

i.e.

$$\frac{4NA^2}{V} > \frac{C_e}{\dfrac{K}{A}\left(\dfrac{C_p}{2} + C_L\right)} \tag{12.31}$$

We have defined

$$\frac{4NA^2}{V} = \lambda_J$$

as the minimum jack stiffness and we also define

$$\left[\dfrac{C_e}{\left[\dfrac{C_p}{2A}\right]}\right]$$

as the valve stiffness λ_v.

$$\therefore \quad \lambda_J > \lambda_v \tag{12.32}$$

when $C_L = 0$, i.e. no leakage is present and say $K = 1$. An initial design criterion is that

$$\lambda_J \text{ is between 2 and } 4\lambda_v$$

(Questions on this section will be incorporated at the end of Chapter 13). *NB*: Sometimes C_p is defined as

$$\frac{\partial Q}{\partial P_L}$$

Appendix 12.1: The perturbation method

Taylors theorem states that if

$$Y = f(X$$

i.e. $Y_i = f(X_i)$ (at a point i on the curve)

for a small change of x in X_i the change in Y_i to $(Y_i + y)$ is given by

$$Y_i + y = f(X_i + x) = f(X_i) + xf'(X_i) + \frac{x_2 f''(X_i)}{2} + \ldots$$

For small perturbation analysis we only consider the first two terms, that is

$$y = xf'(X_1)$$

For two variables i.e. $Y = f(XZ)$

$$y = x \frac{\partial f}{\partial X}\bigg|_i + z \frac{\partial f}{\partial Z}\bigg|_i \quad (|_i - \text{represents mean position (i)})$$

(This is the basis of *linear pertubation* analysis and we must take care at $X \to 0$, for example, to comply with the mathematical limitations of this method). (Exercises are included in those for Chapter 13).

13
Hydraulic Servos: The Compressible Analysis—Part 2

Section 13.1 can be omitted for those without a knowledge of state space analysis. [54]

13.1. The mathematical model in state space form

The linearized equations derived in chapter 12, lead to the two basic relationships for the case when $C_L = 0$ of

$$\frac{C_e}{KA}(\bar{\theta}_1 - \bar{\theta}_0) = \frac{V}{4NA}\frac{dp_L}{dt} + \frac{d\bar{\theta}_0}{dt} + \frac{C_p}{2A}\,'p_L \tag{13.1}$$

and

$$p_L = \left|\frac{M}{A}\frac{d^2\bar{\theta}_0}{dt^2}\right. \tag{13.2}$$

If state variables $x_1(t)$, $x_2(t)$, $x_3(t)$ are introduced together with the input variable u in accordance with the definitions

$$x_1 = P_L \tag{13.3}$$

$$x_2 = \bar{\theta}_0 \tag{13.4}$$

$$x_3 = \frac{d\bar{\theta}_0}{dt} \tag{13.5}$$

and

$$u = \bar{\theta}_i \tag{13.6}$$

then equations (13.1) and (13.2) can be written in the form

$$\frac{V}{4NA}\frac{dx_1}{dt} + \frac{dx_2}{dt} + \frac{C_p}{2A}x_1 = \frac{C_e}{AK}u - \frac{C_e}{AK}x_2 \tag{13.7}$$

and

$$x_1 = \frac{M}{A}\frac{dx_3}{dt} \tag{13.8}$$

93

together with the augmented equation

$$\frac{dx_2}{dt} = x_3 \tag{13.9}$$

In vector matrix form

$$
\begin{bmatrix}
\dfrac{V}{4NA} & 1 & 0 \\[2mm]
0 & 0 & \dfrac{M}{A} \\[2mm]
0 & 1 & 0
\end{bmatrix}
\begin{bmatrix} x_1 \\ x_2 \\ x_3 \end{bmatrix}
=
\begin{bmatrix}
\dfrac{-C_\text{p}}{2A} & \dfrac{-C_\text{E}}{AK} & 0 \\[2mm]
1 & 0 & 0 \\[2mm]
0 & 0 & 1
\end{bmatrix}
\begin{bmatrix} x_1 \\ x_2 \\ x_3 \end{bmatrix}
+
\begin{bmatrix} \dfrac{-C_\text{e}}{AK} \\[2mm] 0 \\[2mm] 0 \end{bmatrix} u \tag{13.10}
$$

or
$$A_2 \mathbf{x} = A_1 \mathbf{x} + \mathbf{b}_1 u \tag{13.11}$$

Where the state vector $\mathbf{x}$ is given by the equation

$$
\mathbf{x} = \begin{bmatrix} x_1 \\ x_2 \\ x_3 \end{bmatrix} = \begin{bmatrix} p_\text{L} \\ \bar{\theta}_0 \\ \dot{\bar{\theta}}_0 \end{bmatrix} \tag{13.11a}
$$

If equation (13.11) is premultiplied by A_2^{-1} then the resulting vector matrix differential equation of state is in the standard form

$$x(t) = A'\mathbf{x}(t) + bu(t) \tag{13.12}$$

where

$$
A' = A_2^{-1} A_1 =
\begin{bmatrix}
\dfrac{-2NC_\text{p}}{V} & \dfrac{-4NC_\text{e}}{VK} & \dfrac{-4NA}{V} \\[3mm]
0 & 0 & 0 \\[3mm]
\dfrac{A}{M} & 0 & 0
\end{bmatrix} \tag{13.12a}
$$

$$
b = A_2^{-1} b_1 =
\begin{bmatrix}
\dfrac{4NC_\text{e}}{VK} \\[3mm]
0 \\[2mm]
0
\end{bmatrix} \tag{13.13}
$$

and the transfer vector of the system satisfies the expression

$$\mathbf{x}(s) = (sI - A')^{-1} bu(s) = \mathbf{g}(s) u(s)$$

so that

$$\mathbf{g}(s) = \frac{adj(sI - A')}{|sI - A'|}$$

$$= \begin{bmatrix} (2NC_\mathrm{p}/V)s^2/\Delta(s) \\ (4NAC_\mathrm{e}/KMV)/\Delta(s) \\ (4NAC_\mathrm{e}/KMV)s/\Delta(s) \end{bmatrix} \qquad (13.14)$$

where the characteristic equation is

$$\Delta(s) = s^3 + (2NC_\mathrm{p}/V)s^2 + (4NA^2/MV)s + (4NAC_\mathrm{e}/KMV) = 0 \quad (13.5)$$

$$= s^3 + a_2 s^2 + a_1 s + a_0 \qquad (13.16)$$

Thus the three transfer functions of the systems can be written in the form

$$\frac{p_\mathrm{L}(s)}{\bar{\theta}_\mathrm{i}(s)} = \frac{(A/M)a_0 s^2}{(s^3 + a_2 s^2 + a_1 s + a_0)} \qquad (13.17)$$

$$\frac{\bar{\theta}_0(s)}{\bar{\theta}_\mathrm{i}(s)} = \frac{a_0}{(s^3 + a_2 s^2 + a_1 s + a_0)} \qquad (13.18)$$

and

$$\frac{\dot{\bar{\theta}}_0(s)}{\bar{\theta}_1(s)} = \frac{a_0 s}{(s^3 + a_2 s^2 + a_1 s + a_0)} \qquad (13.19)$$

This third-order system will be stable if

$$a_2 > 0 \qquad (13.20)$$

$$a_1 > 0 \qquad (13.21)$$

and

$$a_2 a_1 > a_0 > 0 \qquad (13.22)$$

And in terms of the parameters defined in equations the system will be stable if

$$\frac{2NC_\mathrm{p}}{V} \cdot \frac{4NA^2}{MV} > \frac{4NAC_\mathrm{e}}{MVK} \qquad (13.23)$$

or

$$\frac{4NA^2}{V} > \frac{\left(\dfrac{C_\mathrm{e}}{K}\right)}{\left(\dfrac{C_\mathrm{p}}{2A}\right)} \qquad (13.24)$$

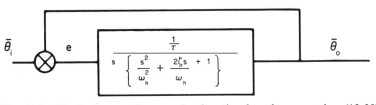

Fig. 13.1   Block diagram for transfer function based on equation (12.29)

i.e. $\lambda_J > \lambda_V$ (13.25)

as derived in equation (12.28), Chapter 12.

The block diagram representing the system is shown in Fig. 13.1 (equations (13.18) and (12.29).

13.2 Introduction of a feedback lever

Let us now examine modifications caused by the introduction of the feedback lever as shown (Fig. 13.2) [3]. The valve opening is proportional to x, where

$$x = \frac{L_2}{L} \cdot y_1 - \frac{L_1}{L} y_0 \tag{13.26}$$

$$(\text{as } L_1 + L_2 = L) \tag{13.27}$$

We can express the error as the difference between the desired and actual output, i.e.

$$e = \frac{Lx}{L_1} = \frac{L_2 y_i}{L_1} - y_0 \; (= \bar{\theta}_i - \bar{\theta}_0) \tag{13.28}$$

This means that

$$K = \frac{L}{L_1} \tag{13.29}$$

and the time constant

$$\tau = \frac{AL}{L_1 C_e} \tag{13.30}$$

The inequality now becomes

$$\frac{4NA^2}{V} > \frac{C_e L_1}{\dfrac{C_p}{2A}} \tag{13.31}$$

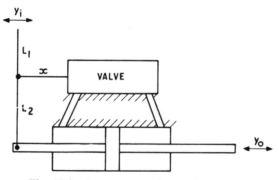

Fig. 13.2   Lever system for section 13.2

VALVE OPENING $= x = y_i - y_o = e$

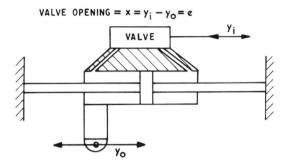

Fig. 13.3 $K = 1$: valve body fixed to cylinder

Often units are designed so that $K = 1$, and one is shown in Fig. 13.3. Here the piston rod is fixed and the valve body is attached to the cylinder which itself moves against the load. The resultant valve opening is then given, in terms of small displacements by

$$x = y_i - y_0 = e = \bar{\theta}_i - \bar{\theta}_0 \qquad (13.32)$$

The analysis of a three-way valve and jack is given in Stringer [3].

13.3 Example

A valve-controlled jack horizontally positions a weight of 2225 N within a maximum stroke of ± 7.6 cm about the mid-position. With supply pressure of 54 bar the maximum acceleration at zero velocity is 12.7 m/s^2 with this load of 2225 N. Calculate the undamped natural frequency of the system. (Ram efficiency = 100%).

With the spool centrally placed, a small spool displacement of 0.025 mm causes a flow of 4.42 cm^3/s. In order to stabilize the system a viscous leak is to be provided across the ram piston in the form of a hole drilled through the piston which is 3.8 cm thick. If the hydraulic fluid has a kinematic viscosity of 30 centistokes (cS) and specific gravity of 0.87 calculate the minimum diameter of hole required for stability ($C_p \to 0$). $N = 18 \times 10^8$ N/m^2. ($K = 1$)

NB: For the laminar flow of a fluid through a pipe, the flow rate is given by

$$Q = \frac{\pi}{128} \frac{\Delta_p}{L} \frac{d^4}{\mu} \text{ i.e. } C_L = \frac{\pi}{128} \cdot \frac{1}{L} \cdot \frac{d^4}{\mu}$$

where μ is the absolute viscosity of the fluid, L is the length and d the diameter of the pipe; Δ_p is the pressure drop. (If an orifice is used we differentiate the appropriate formula).

Maximum acceleration occurs when velocity is zero, that is,

$$p_2 = 0$$

$$P_1 = P_s$$

$$\therefore \quad P_sA = M\frac{d^2x}{dt^2}$$

i.e.

$$A = \frac{2225}{9.81} \times \frac{12.7 \times 10^4}{54 \times 10^5} = 5.33 \text{ cm}^2$$

(diameter = 2.6 cm)

$$V = 5.33 \times 2 \times 7.6 = 81 \text{ cm}^3$$

For $x = 0.0025$ cm then $Q = 4.42 \text{ cm}^3/\text{s}$

$$\therefore \quad C_e = \frac{\partial O}{\partial X} = \frac{4.42}{0.0025} = 1.77 \times 10^3 \text{ cm}^2/\text{s}$$

For stability $\dfrac{\dfrac{C_e}{C_L}}{A} < \dfrac{4NA^2}{V}$ for $C_p = 0$

$$\therefore \quad C_L > \frac{1.77 \times 10^3 \times 10^{-4} \times 81 \times 10^{-6}}{4 \times 18 \times 10^8 \times 5.33 \times 10^{-4}}$$

i.e.

$$C_L > 3.75 \times 10^{-11}$$

also

$$C_L = \frac{\pi}{128} \times \frac{d^4}{30 \times 10^{-6} \times 870} \times \frac{10^2}{3.8} = 3.75 \times 10^{-11}$$

$$\therefore \quad d > 1.1 \text{ mm}$$

13.4 Exercises

(A) Self-assessment

1. Consider the assumptions in Section 12.2. Examine the open loop transfer function and other equations. Assess the validity of the assumptions.
2. Will a constant load affect the conclusions?
3. Consider the effect of external viscous friction.
4. 'Coulomb friction adds a non-linearity'—consider this statement.
5. Feedback levers are nearly always an integral part of the controls. Can you consider alternative ways in which they can be connected.

(B) Numerical

1. A four-way valve with 'full' periphery annular ports has a 6 mm diameter spool and it may be assumed that the spool lands fully cover the valve ports in the zero or midposition. Estimate the flow rate through one port when the pressure drop across it is 70 bar for every millimetre spool displacement. ($C_q = 0.6$) (1.44 1/s/mm).
2. What would be the flow gain C_e of the above valve if it was used as a part

of a servo system having an oil supply pressure (a) 140 bar, (b) 210 bar?
((a) 1.44, (b) 1.76 (1/s/mm))

3. A three-way spool valve with half the annular periphery of the valve port blocked off and spool diameter 9 mm is used in a system supplied with oil at 120 bar pressure. The 'half-area' piston has areas of 0.004 m^2 and 0.002 m^2 and a maximum required velocity of 0.3 m/s. Estimate the maximum spool displacement required. (about 1.2 mm)

4. In a 240 bar servo-system employing a four-way valve, valve underlap is used to assist in damping system oscillations. The valve has a 4 mm diameter spool, full periphery ports and nominal underlap of 0.0127 mm. Estimate the pressure-flow coefficient C_p for the valve. $(13.4 \times 10^8 \text{ m}^3/\text{s/bar})$

5. A simple valve controlled servo-system consists of a closed centre metering valve and a tailrod actuator. There is an equal length feedback lever connecting input, valve and jack. Each valve control port orifice has an area of 1.25 mm^2 per mm of spool displacement. The oil density is 700 kg/m^3 and the supply pressure is 70 bar. The actuator piston area is 32.25 cm^2. Show that the time constant is 0.0064.s for $C_q = 0.8$.

6. The block diagram shown in Fig. 13.4 represents a flying control unit comprising a four-way valve and jack with a through piston rod connected together by a feedback lever. Determine an expression for ω_h when the jack piston is at the mid-stroke position. Hence calculate the change required in the value of L to maintain the same value of ζ when the fluid in the system is changed from one having a bulk modulus of 15×10^8 to one of 10×10^8 N/m^2 (it will be reduced by 20%).

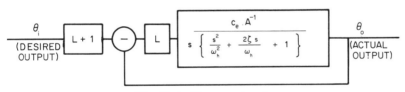

Fig. 13.4   For question 6

14
Factors Affecting Stability

14.1 Valve characteristics

As we have previously seen (in Chapter 10) it is possible for the valve lands to have dimensions relative to the port widths which provide different characteristics that we label (i) zero lap, (ii) overlap, (iii) underlap [14]. In all cases we consider the valve openings on both sides of the jack to be symmetrical and for the purposes of comparison we assume a linear opening valve, in other words where the area is proportional to spool displacement (X). The relative dimensions are illustrated in Fig. 14.1.

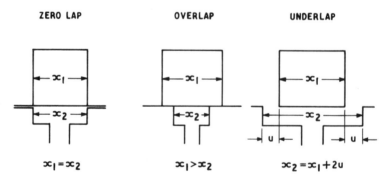

Fig. 14.1   Valve laps

14.2 A comparison of the three cases

(K = 1, see Chapters 12 and 13)[3], [14].

14.2.1 Zero lap

Here

$$Q_1 = CX(P_s - P_1)^{1/2} : (C \text{ is a constant}) \tag{14.1}$$

100

where C subsumes all the various constants in the orifice equation.

$$\therefore \quad C_e = \frac{\partial Q_1}{\partial X}\bigg|_i = C(P_s - P_1)^{\frac{1}{2}} \tag{14.2a}$$

$$C_p = \frac{-\partial Q}{\partial p_1}\bigg|_i = \frac{CX}{2(P_s - P_1)^{\frac{1}{2}}} \tag{14.2b}$$

$$\therefore \quad \lambda_v = \frac{C_e}{\dfrac{C_p}{2A}} = 4A\frac{(P_s - P_1)}{X} \tag{14.3}$$

and as $X \to 0$

$$\lambda_v \to \infty \quad \therefore \quad \lambda_J \not> \lambda_v$$

i.e. the system is unstable, therefore added damping is required!

14.2.2   Overlap

Here the condition is similar to that above. Nothing happens while moving in the 'overlap' region and drifting of the control can occur within these dimensions. However, as soon as the valve begins to open we have a situation similar to that in 14.2.1 and instability can occur.

14.2.3   Underlap

$$Q_1 = C(u + X)(P_s - P_1)^{\frac{1}{2}} - C(u - X)(P_1)^{\frac{1}{2}} \tag{14.4}$$

$$\therefore \quad C_e = C(P_s - P_1)^{\frac{1}{2}} + C(P_1)^{\frac{1}{2}} \tag{14.5}$$

$$C_p = \frac{C(u + X)}{2(P_s - P_1)^{\frac{1}{2}}} + \frac{C(u - X)}{2(P_1)^{\frac{1}{2}}}$$

Let us examine the condition as $P_1 = P_2$

$$\text{i.e. } P_1 = P_2 = \frac{P_s}{2}$$

If we let $X \to 0$ then

$$\frac{C_e}{\dfrac{C_p}{2A}} = \frac{2AP_s}{u} \tag{14.7}$$

For stability

$$\frac{4NA^2}{V} > \frac{2AP_s}{u} \tag{14.8}$$

i.e.

$$u > \frac{VP_s}{2NA} \tag{14.9}$$

Theoretically it is possible for underlap to provide stability. Let us now put some numbers into this inequality

say $N = 18 \times 10^8$ bar, $P_s = 300$ bar, $V = 450$ cm^3 and $A = 60$ cm^2

Then

$$u > \frac{1}{2} \times 450 \times \frac{3 \times 10^7}{18 \times 10^8} \times \frac{1}{60} \text{ cm}$$

or $\qquad u > 0.6$ cm.

An underlap dimension of this magnitude would be impossible. Underlap on its own will not provide adequate stability.

14.3   Valve flow forces

When considering the normal four land valves illustrated in Fig. 14.2(a) there is net zero force [3], [43], [44] acting on the valve spool as the pressure pushes equally on lands (1) and (2) in opposite directions when the valve is closed. What happens, however, when it is opened?

Due to the increase in velocity of the fluid as it approaches the metering lands when flowing in and out of the valve the pressure is reduced and a resultant force arises which tends to close the valve. For a valve with two metering orifices this closing force (F) is calculated approximately by using the momentum equation leading to

$$F = [6.6Q(\Delta p)^{\frac{1}{2}}] \text{ N} \qquad (14.10)$$

where Q is in litres/s and Δp in bar for a fluid of specific gravity $= 0.85$. As the valve is opened it tries to close. Unless sufficient force is applied oscillations occur with the valve banging 'open' and 'shut'.

How can we avoid this? Figure 14.2(b) shows one approach. The spool and downstream chamber are shaped to provide an opening force to balance the closing force from the unshaped upstream port. With correct shaping the forces can be made to balance almost exactly in the manner indicated in Fig. 14.2c but the method is expensive and sensitive to small dimensional changes.

14.4   A general description of how instability can occur

This is based on Fig. 14.3 [1]. The major factors affecting stability are the sensitivity and response together with the inertia, resilience, and backlash in the system. Detail design care is therefore essential.

'The sensitivity is determined by the ability to detect small errors and be moved by small forces. Unless the output member has a small mass stability is difficult to maintain. The difficulty is caused by having to bring the output mass to rest without causing a reversal of the control action. Referring to Fig. 14.3, when the input is brought to rest the valve will

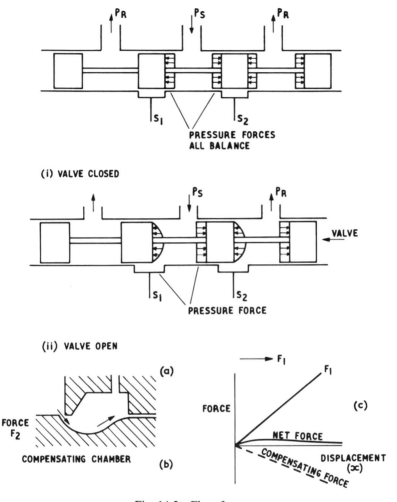

Fig. 14.2   Flow forces

close and the output mass have to be brought to rest by the hydraulic jack. A high retarding force is necessary and this causes axial deflecton at the jack mounting due to possible aircraft structure distortion and what is more important by the compressibility of the fluid. The feedback lever AB moves and the control valve is opened in the opposite direction as the fluid compresses. This can reverse the control action opening the wrong port to pressure within the valve causing the jack to go into 'reverse' producing an oscillatory motion. Reducing the response rate helps to alleviate this' [1].

As structural deflections also cause this to happen it is essential to mount the jack in such a way that these do not feed back to exacerbate the condition.

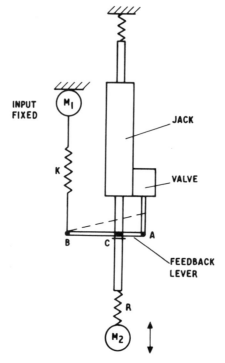

Fig. 14.3   General stiffness affecting stability

14.5   Some general conclusions

To alleviate and cure the tendency to instability [13].

1. Strict attention must be given to detail design of valve and levers.
2. Fluid compressibility must be reduced, in other words, increase N:
    (a) use a rotary motor if possible;
    (b) use a large jack if possible;
    (c) eliminate entrained air.
3. Avoid structural feedback.
4. Eliminate backlash in all joints (use preloaded bearings).
5. Provide a viscous leak across the piston.
6. Compensate for valve flow forces.
7. Add velocity feedback (see Chapter 18).
8. Add derivative of the error (see Chapter 18).
9. Add pressure feedback (see Chapter 18).
10. Add phase advance.

NB: *In general instability is reduced if the natural frequencies of the input and output circuits are made much larger than the natural frequency of the jack* [13].

Because the use of servo-controls requires very little effort, it is necessary to introduce 'artificial' feel [1] into the system to enable the pilot to maintain some form of control over his reactions when moving the various surfaces.

14.6   Exercises

(A) Self-assessment

1. Consider Fig. 14.3 and note how the stiffness of jack and structure can account for servo-instability.
2. Why should backlash in the joints affect stability?
3. Why does the viscous leak across the piston not lead to as high a power loss as 'underlap'?
4. What disadvantage does the 'viscous leak' have?

(B) Numerical

1. A servomechanism incorporating a hydraulic jack is known to be stable if

$$\zeta > 1/(2\omega_h\tau),$$

where ζ is the non-dimensional damping factor, τ is the time constant of the system, and ω_h is the undamped natural frequency of the isolated jack system.

   Such a hydraulic servomechanism consists of a four-way valve which controls the flow rate into the cylinder of a hydraulic jack with a through piston rod.

   The supply pressure is 200 bar and the 'stalled load' of the actuator, which is assumed to be 100% efficient, is 40 kN. Determine the minimum value of ζ if the stroke is 0.25 m, the mass of the load is 2000 kg, the bulk modulus of the oil is 12.5 kbar, and τ is 5×10^{-3} s; the volume of oil in the lines can be neglected. (Stalled load occurs at zero velocity with the valve open, i.e. $p_s = p_1$). (0.7)

2. A flying control unit comprises a four-way valve coupled to a hydraulic jack with a through piston rod. Calculate the undamped natural frequency from the data listed below:

   (a) Stalled load — 20 kN (jack efficiency 100%).
   (b) Maximum system pressure — 250 bar.
   (c) Stroke — 0.5 m (the total volume can be based on the stroke and area of the piston).
   (d) Effective bulk modulus of the fluid — 15 kbar.
   (e) Mass of control surface referred to the jack — 50 kg. (69.7).

3. For a four-way valve controlling a hydraulic jack with a through piston rod it is stated that the design criterion used to maintain stability is that

$$\lambda_J > 4\lambda_v$$

where

$$\lambda_J = \frac{4NA^2}{V}$$

and

$$\lambda_v = \frac{2AC_e}{C_p}$$

If the valve is symmetrical and the lands have an underlap to satisfy the condition specified above discuss the feasibility of this method of stabilizing the unit. The appropriate numerical data are as follows:

A: area of piston in contact with the fluid — 15 cm^2
N: effective bulk modulus — 18 kbar
V: total enclosed volume between valve and jack — 450 cm^3.

The supply pressure is constant at 300 bar. C_e and C_p are the valve 'gain' and 'pressure factor', respectively. The jack is assumed to be 100% efficient and the pressures on each side of the piston are equal. (1 cm not practical)

4. A hydraulic servo consists of a variable-delivery pump and a fixed-displacement rotary motor which drives an inertial load, J. The pump-control mechanism subtracts the actual output position, θ_i, and provides a flow rate, Q_p, which is given by

$$Q_p = A(\theta_i - \theta_0) \text{where } A \text{ is a constant.}$$

Show that, in the absence of external damping, the transfer function, $\bar{\theta}_0(s)/\bar{\theta}_i(s)$, can be expressed in the form

$$\frac{\bar{\theta}_0(s)}{\bar{\theta}_i(s)} = \frac{\dfrac{1}{\tau}}{s^3/\omega_h^2 + 2\zeta s^2/\omega_h + s + \dfrac{1}{\tau}}$$

expressing τ, ω_h, and ζ in terms of the system constants, C_L, the leakage flow-rate per unit pressure difference between the fluid lines; V, the volume contained (one pipe only pressurized), N, the bulk modulus of the fluid; and D, the motor displacement per radian.

Determine the maximum volume, V, for which the system is marginally stable for the case when

$D = 15 \text{ cm}^3/\text{rad}$,
$C_L = 4 \times 10^{-2} \text{ cm}^3/\text{s/bar}$
$C = 1.5 \text{ cm}^3/\text{s/rad}$,
$N = 12 \text{ kbar (0.96 litres)}$

5. Determine the magnitude of the flow reaction force on a four-way spool, diameter 8 mm, pressure drop to tank. 136 bar, opening 0.3 mm. $C_q = 0.8$, $\varrho = 850 \text{ kg/m}^3$. The ports occupy 50% of the circular periphery. (41.4 N)

15
Harmonic Response—
The Bode Diagram

(*NB* Any general text on control theory will serve as an introduction to the contents of this chapter)

15.1 Introduction

The response of the servo to a sinusoidal input can tell us a great deal about its characteristics and performance [3]. It indicates the range of stability and allows an estimate of the 'gain' and 'phase' margins to be made to ensure that these are within acceptable limits. Although many methods such as Nyquist, Root Locus and others are available it is the 'BODE' diagram which has best served the interest of fluid power engineers. It provides an immediate indication of the range of stability and how additive networks can improve this.

Basically, the diagram plots the amplitude ratio in *decibels* against either the *frequency* in *octaves* (or decades) and the *phase margin in degrees*. This is done for the open loop transfer function both diagrammatically and numerically providing a complete picture of the response of the system.

15.2 The frequency response—first and second order systems

The equation connecting the output position and error, in Laplacian form, neglecting load disturbances, can be represented by

$$\frac{\bar{\theta}_0(s)}{e(s)} = \frac{\dfrac{1}{\tau}}{s\left(\dfrac{s^2}{\omega_h^2} + \dfrac{2\zeta s}{\omega_h} + 1\right)} \tag{15.1}$$

This can then be transformed into the *frequency* response

$$\frac{\bar{\theta}_0(j\omega)}{e(j\omega)} = \frac{\dfrac{1}{\tau}}{j\omega\left(\left(\dfrac{j\omega}{\omega_h}\right)^2 + 2\zeta\left(\dfrac{j\omega}{\omega_h}\right) + 1\right)} \tag{15.2}$$

by the substitution of $s = j\omega$ where $j = (-1)^{\frac{1}{2}}$

The first term $\dfrac{1}{j\omega\tau}$ represents an integration

i.e. $$\bar{\theta}_0 = \frac{1}{\tau}\int edt \tag{15.3}$$

If $$e = \hat{e}\sin\omega t \tag{15.4}$$

$$\bar{\theta}_0 = \frac{1}{\omega\tau}\hat{e}\cos\omega t = \frac{1}{\omega\tau}\hat{e}\sin(\omega t - 90°) \tag{15.5}$$

i.e. the amplitude ratio

$$\left|\frac{\bar{\theta}_0}{e}\right| = \frac{1}{\omega\tau} = \frac{1}{\left(\dfrac{\omega}{\omega_h}\right)(\omega_h\tau)} = \frac{1}{r(\omega_h\tau)} \tag{15.6}$$

where $r = \left(\dfrac{\omega}{\omega_h}\right)$

and the phase lag is $90°$. This is shown in Fig. 15.1 where the amplitude ratio is plotted in decibels (db)

i.e. $$20\log_{10}\left|\frac{\bar{\theta}_0}{e}\right|$$

and the horizontal scale is in octaves.

Doubling the frequency doubles the amplitude ratio i.e.

$$\frac{\left|\dfrac{\bar{\theta}_0}{e}\right|_{\omega_1}}{\left|\dfrac{\bar{\theta}_0}{e}\right|_{\omega_2}} = \frac{\dfrac{1}{\omega_1\tau}}{\dfrac{1}{\omega_2\tau}} = \frac{\dfrac{1}{\omega_1}}{\dfrac{1}{2\omega_2}} = 2 \tag{15.7}$$

and $20\log_{10}2 = 6$ db $\tag{15.8}$

$\therefore$ The slope is 6 db per octave

0 db occurs when $20\log_{10}1 = 0$, i.e. when $\omega = \dfrac{1}{\tau}$.

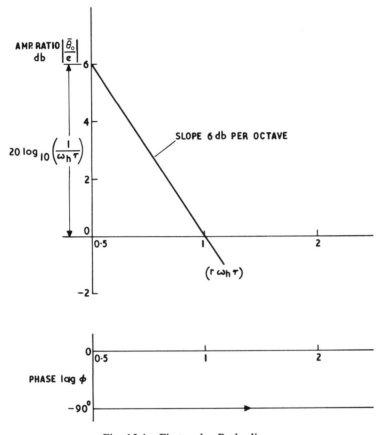

Fig. 15.1 First-order Bode diagram

The second term in equation (15.2) is of the form

$$\frac{1}{\left(\dfrac{j\omega}{\omega_h}\right)^2 + 2\left(\dfrac{j\omega}{\omega_h}\right)\zeta + 1}$$

(This is a mass, spring-damper system). For a sinusoidal input

$$\left|\frac{\bar{\theta}_0}{e}\right| = \frac{1}{\{(1 - r^2)^2 + 4\zeta^2 r^2\}^{1/2}} \sin(\omega t - \phi) \qquad (15.9)$$

$$\text{and } \tan \phi = \frac{2\zeta r}{1 - r^2} \qquad (15.10)$$

The asymptotic line for this is a fall of 12 db per octave, for example,

r is small

$$\tan \phi = 2\zeta r$$

$$\phi = \tan^{-1} 2\zeta r \text{ and } \phi \to 0$$

$$\left|\frac{\bar{\theta}_0}{e}\right| = \frac{1}{(1 - 4\zeta^2 r^2)^{\frac{1}{2}}} = 1 \tag{15.11}$$

$$\therefore \quad 20 \log_{10} 1 = 0$$

At high values $\tan \phi = \dfrac{-2\zeta}{r} = -0$

$$\text{i.e. } \phi \to 180^\circ \tag{15.12}$$

$$\left|\frac{\bar{\theta}_0}{e}\right| = \left[\frac{1}{[-r^2]^2}\right]^{\frac{1}{2}} = +\frac{1}{r^2} \tag{15.13}$$

Double ω and

$$\frac{\left|\dfrac{\bar{\theta}_0}{e}\right|_{\omega_1}}{\left|\dfrac{\bar{\theta}_0}{e}\right|} \quad \frac{\dfrac{1}{\omega_1} \cdot 2}{\dfrac{1}{4\omega_1} \cdot 2} = 4 \tag{15.14}$$

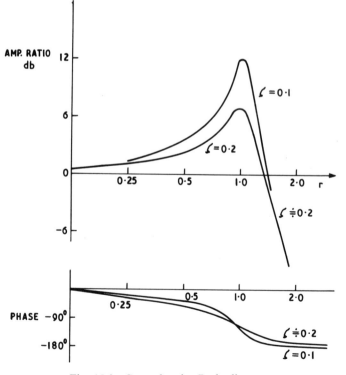

Fig. 15.2　Second-order Bode diagram

$$20 \log_{10} 4 = 12 \qquad (15.15)$$

when $r = 1$ $\tan \phi = \infty$ $\therefore$ $\phi = 90°$

15.3 The complete Bode diagram

The combination of the two elements is now shown in Fig. 15.3 with the important points clearly defined. It must be noted that the phase and gain margins usually required are:

1. Gain margin > 6 db (where the phase lag = $180°$)
2. Phase margin > $45°$ (where the 0 db axis is crossed)
   i.e. Phase angle = -135

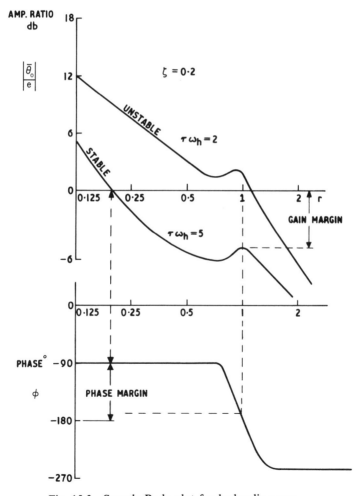

Fig. 15.3 Sample Bode plot for hydraulic servo

(The closed loop amplitude ratio = 3 db)
From equation (15.2) the amplitude ratio is given by

$$\left|\frac{\theta_0}{e}\right| db = \left|\left|\frac{1}{r(\tau\omega_h)}\right| - r - ((1 - r^2)^2 + (2\zeta r)^2)^{\frac{1}{2}}\right| db \qquad (15.16)$$

The acceptable stability is at the point when $r = 1$
For this

$$2\zeta > \frac{1}{\tau\omega_h}$$

or

$$\zeta > \frac{1}{2\tau\omega_h}$$

which was proven in Chapters 12, 13 and 14.
Assymetrical jack volumes are covered in Martin [45].

15.4 Exercises

(A) Self-assessment

1. At what value of (r) does the first-order system cross the axis?
2. What is the open loop amplitude ratio for a gain margin of 6 db?
3. What are the gain and phase margins?
4. Can you relate the closed loop amplitude ratio to the 'gain margin'.

(B) Numerical

1. Devise a computer program and test it to plot the frequency response on a BODE diagram of the amplitude ratio. Try $\tau = 0.01$ to 0.05, $\omega_h = 200$ to 90 with $\zeta = 0.25$ and 0.1.
2. Tabulate the *gain margin* for question 1.
3. An aircraft hydraulic servo-actuator is geared to a control surface. The moment of inertia of the control surface about its hinge line is J; the inertia of the other moving parts may be neglected. The actuator is controlled by a four-way valve and displaces a volume D of fluid per radian rotation of the control surface. The volumes of fluid contained between the actuator and the valve are $V/2$ either side of the actuator. The effective bulk modulus of the fluid is N. The servo is designed so that the flow from the valve to the unloaded actuator gives it a velocity which corresponds to an angular velocity of the control surface given by the formula,

where
$$\frac{d\theta_0}{dt} = \frac{(\theta_i - \theta_0)}{\tau},$$

θ_0 is the angular displacement of the control surface,
θ_i is the control surface displacement called for by the input, and
τ is a constant.

(a) Show that the open-loop undamped natural frequency of the control surface supported on the resilience of the contained oil column is

$$W_h = \left(\frac{4D^2N}{VJ}\right)^{\frac{1}{2}} \text{ rad/s}$$

Hence determine this frequency when

$D = 120 \text{ cm}^3$ $N = 7 \text{ kbar}$
$V = 200 \text{ cm}^3$ $J = 0.28 \text{ kg} \cdot \text{m}^2$ (848 rads/s)

(b) For this servo, indicate the range within which the maximum value of τ giving adequate stability may be expected to lie when $0.1 < \zeta < 0.2$.

(c) Sketch the BODE amplitude ratio diagram

(d) if $V = \alpha D$, show that for stability.

$$V = \left(\frac{\alpha w_h^2}{2}\right)\left(\frac{J}{N}\right)\left(0.2 < \frac{1}{\tau w_h} < 0.4\right).$$

16
Electrohydraulic Servos

16.1 Introduction

A brief description of an electrohydraulic servo-valve [3], [13], [46] together with an illustration was given in Chapter 8 (Fig. 8.3, p. 62) and included an analysis of how the flapper nozzle valve is used as a pilot stage to move the main spool. This, however, is only one of a diversity of designs which have been used as servo-valves for aircraft applications. They are all mainly single or two-stage units where the first stage is driven by an electromagnetic torque motor itself controlled by the error signal between the desired and actual output in the form of a voltage. This when applied to the coils is converted into a differential current (I), resulting in an electromagnetic force so that the flow is given by

$$Q = KI(P_s - P_L)^{1/2} \ (K \text{ is a constant}) \tag{16.1}$$

Because the design of such valves is highly specialized the reader is given many references which he can study as the need arises. Meanwhile this section deals with the application of such valves to the drive of a linear actuator, the voltage signals being taken from potentiometers at the output and input.

16.2 Valve characteristics

The response of a typical valve is shown in Fig. 16.1 [3]. At its simplest the response approximates to a 'first-order' type dependent only on a proportionality constant at the frequency normally encountered (using small perturbations); (we denote perturbation current as I).

$$\frac{x}{I} = \frac{K_I}{1 + \tau s} \doteq K_I \tag{16.2}$$

in other words, the displacement of the main spool is a function of the input current. This itself is a linear function of the differential voltage derived from the potentiometers at the input and output positions and the resistance of the coils in the torque motor.

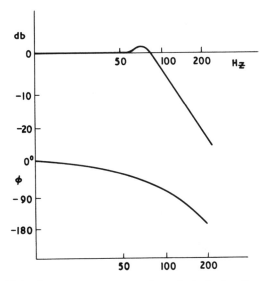

Fig. 16.1 Frequency response of an electrohydraulic valve

16.3 The linearized analysis of the complete electrohydraulic servo-unit

The diagram for the model is illustrated in Fig. 16.2. The valve spool displacement and output of the jack are connected by the transfer function

$$\frac{y_0}{x} = \frac{\dfrac{C_e}{A}}{s\left[\dfrac{s^2}{\omega_h^2} + \dfrac{2\zeta s}{\omega_h} + 1\right]} \tag{16.3}$$

derived in a similar manner to Chapters 12 and 13.

The displacement of the spool in an electrohydraulic valve is a function of

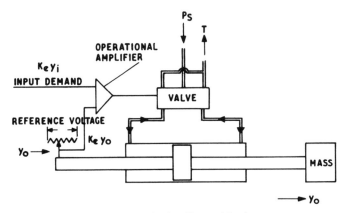

Fig. 16.2 Electrohydraulic positioning system

the input differential current (this produces a magnetic field), i.e.

$$\frac{x}{I} = K_I \text{ (meter/amp)}$$

∴ The overall transfer function for the valve is

$$\frac{y_0}{I} = \frac{\dfrac{K_I C_e}{A}}{s\left(\dfrac{s^2}{\omega_h^2} + \dfrac{2\zeta}{\omega_h}s + 1\right)} \text{ (metre/amp)} \tag{16.4}$$

When we considered the output position of the piston the displacement was converted into a proportional reference voltage:

$$\frac{V_0}{y_0} = K_e \text{ (volts/metre)} \tag{16.5}$$

The input is also taken from a similar potentiometer and therefore the input (= desired output) is given by

$$\frac{V_i}{y_i} = K_e \tag{16.6}$$

The error voltage is then connected to the positional error e

$$e_v = V_i - V_0 = K_e(y_i - y_0) = K_e e \tag{16.7}$$

This is small and has to be amplified to provide a larger voltage Ω to control the torque motor

$$\frac{\Omega}{e_v} = K_d \text{ (volts/volts)} \tag{16.8}$$

which is now applied to the torque motor coils to produce a current I where

$$\frac{\Omega}{I} = \frac{1}{K_R} \text{ (ohm)} \tag{16.9}$$

The total transfer function is given by the product of (16.1) to (16.9).

$$\frac{V_0}{e_v} = \frac{K_e K_d K_R \left[\dfrac{K_I C_e}{A}\right]}{s\left(\dfrac{s^2}{\omega_h^2} + \dfrac{2\zeta s}{\omega_h} + 1\right)}$$

We let $\quad K_e K_d K_e \left[\dfrac{K_I C_e}{A}\right] = \dfrac{K_{qe}}{A}$

16.4 Block diagram and general implications

This is shown in Fig. 16.3 [3], [13], [46]. It can be inferred that

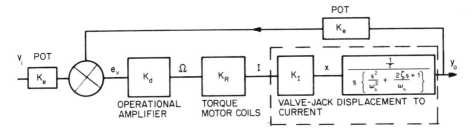

Fig. 16.3 Block diagram of electrohydraulic system

1. This type of unit is mainly used as a position servo.

2. There is a limitation on the frequency response.

3. It is assumed that the electrical response of the valve torque motor is instantaneous.

4. (C_e/A) should be taken as the basis for ζ.
Merritt [13] suggests $0.1 < \zeta < 0.2$ and (K_{qe}/A) between the limits of 20 and 40% of ω_h.

5. The rating of the value in $(m^3/s/(ma))$ must be multiplied by $(P_s/\Delta_p)^{\frac{1}{2}}$ to give $K_I C_e$. Do not select too large a valve!
(Δ_p is the pressure drop at which the valve is tested.)

16.5 Some important features of electrohydraulic servo-valves

The following are important features [13], [46], [47]:

1. Flow range.
2. Maximum supply pressure.
3. Rated input signal (normally 8–200 ma).
4. Typical frequency response at $90°$ phase lag.
5. Frequency for -3 db.
6. Weight.
7. Flow gain and linearity.
8. Null leakage — that is, leakage in neutral position.
9. Pressure gain.
10. Null shift (drift at zero position).
11. Resolution (less than 1.5% without dither*).
12. Coil resistance and inductance.
13. Hysteresis ($< 3\%$ rated current).
14. Load error (*most important*).

NB *Dither is a high frequency signal superimposed on other signals to prevent stiction.

16.6 Example

Assume

$$\omega_h = 120 \text{ rad/s}$$

Coil resistance $= 400$ ohm

$$\Delta p = 60 \text{ bar for a flow gain of } 0.04 \text{ litres/s/ma}$$
$$\text{i.e. } [K_I C_e]$$

$$A = 16.0 \text{ cm}^2$$

Say
$$\frac{K_{qe}}{A} = 33\% \text{ of } \omega_h$$

i.e.
$$\frac{K_{qe}}{A} = 40(1/s) \quad \text{or} \quad K_{qe} = 40 \times 16.0 \times 10^{-4}$$

If supply pressure is $p_s = 200$ bar

$$\therefore \quad (K_I C_e), \text{ the maximum flow gain} = 0.04 \times \left(\frac{200}{60}\right)^{\frac{1}{2}} \times 10^3$$
$$= 73 \text{ litres/s/amp}$$

Say
$$K_R = \frac{1}{400} \text{ 1/ohm}$$

$$\therefore \quad K_e K_d = \frac{K_{qe}}{K_R(K_I C_e)} = \frac{40 \times 16 \times 10^{-4} \times 400}{73 \times 10^{-3}}$$
$$= 350 \text{ volt/metre}$$

If we use a transducer with $K_e = 0.04$ volt/mm then the amplifier gain

$$K_d = \frac{315}{0.04 \times 10^3} = 8$$

16.7 Exercises

(A) Self-assessment

1. Sketch the curves of $Q \sim I$ at constant ΔP and $Q \sim \Delta P$ at constant I.
2. What sort of limitation does the valve impose on the system?
3. What is the effect of hystereses on the $Q \sim I$ curve.
4. How does the gain change with load pressure change?
5. Find out what 'fly by wire' really means and how it differs from previously electrically controlled systems.

(B) Numerical

1. Find the time constant for the open loop transfer function for an electrohydraulic servo (no compressibility) for the data given.

   (a) Control surface transducers 10 mV/degree of control surface
   (b) Differential amplifier
      Output = 0.25 (input–output transducer) ma.
   (c) Servo valve
      25 ma input — flow 125 cm^3/s
   (d) Jack: tailrod: bore 5 cm
                rod 2 cm
   (e) Linkage between jack and control surface
      1 cm jack movement = 9° control surface (0.147 s)

2. An electrohydraulic valve drives a jack moving a surface of mass 200 kg. The amplifier gain is 7 volts/volt. The valve flow gain at one-third system pressure drop across the valve is 0.006 litre/s for ever milliamp (ma). K_R is 2×10^{-3} amp/volt. The total trapped volume is 0.3 litres and the area of the tailrod jack is 5 cm^2. $N = 12 \times 10^8$ N/m^2. Calculate the hydraulic natural frequency and estimate the potentiometer constant K_e. (assume $[K_{qe}/A] = 36$ s^{-1}). (141 rad/s, $\simeq 1.0$ volts/mm)

3. In a valve–jack servomechanism operating about a particular position, the characteristic of the valve may be represented as

$$\delta Q = K_Q \, \delta i - K_c \, \delta P$$

where Q is the valve flow in m^3/s, i is the electric current to the valve in mA, and P is the pressure drop across the jack in N/m^2.

   (a) A servomechanism of this type with an effective jack area of 0.0015 m^2 was tested with different jack loads and current flows to the valve. Some of the results are given in the table below.

Jack speeds in m/s

Current to valve (mA)	Jack load 14000 N	Jack load 15000 N
9.5	0.0697	0.0633
10.5	0.0770	0.0700

Use these results to estimate values for K_Q and K_c and state the operating position about which these values are effective.

   (b) Explain the physical reasons for K_c appearing in the equation and discuss how the value of K_c is likely to change at other operating positions for which the valve current is less than the values given in the table.
   (c) Discuss the effect of K_c on the performance of the servomechanism.

17
The Choice of Components

17.1 Introduction

Loads in real systems are difficult to specify with any exactitude [13],[14]. Theoretically the velocity and acceleration together with the load should be stated both as a function of time and displacement of the actuator. This would enable the designer to plot a 'load locus' to which the hydraulic system could be matched. The realistic way is to specify a limited number of load cases even if this means tabulating the external force, inertia and acceleration at various parts of the stroke plus the external stiffness of the structure. Mainly it is essential to discover and specify the load condition corresponding to maximum power.

Accuracy and the elimination of errors due to input position, velocity and acceleration, drift and external disturbances plus those due to non-linearities are essential. The provision of adequate stability is a prime necessity. One of the usual criteria in the past selection of servo-valves has been to rely on the concept of 'maximum power transmission' and this will now be dealt with.

17.2 Maximum power transmission

Figure 17.1 outlines the diagram used for the argument. Let us assume that at maximum opening of the valve ($P_R = 0$)

$$Q_1 = C_1(P_s - P_1)^{1/2} = Q_0 = C_1(P_2)^{1/2}(C_1 \text{ is a constant}) \quad (17.1)$$

$$P_1 - P_2 = \frac{\text{load}}{\text{area}} = P_L \quad (17.2)$$

$$P_s - P_1 = P_2 \quad (17.3)$$

$$P_2 = \frac{P_s - P_L}{2} \quad (17.4)$$

$$\therefore \quad Q_1 = C_2(P_s - P_L)^{1/2} \ (C_2 \text{ is another constant}) \quad (17.5)$$

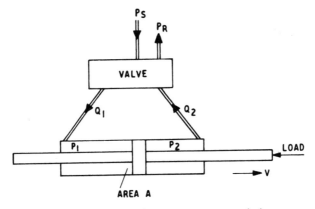

Fig. 17.1 For maximum power transmission

The power delivered to the load

$$W = P_L Q_1 = C_2 P_L (P_s - P_L)^{\frac{1}{2}} \tag{17.6}$$

$$\frac{dW}{dP_L} = -\frac{P_L}{2(P_s - P_L)^{\frac{1}{2}}} = 0 \text{ for maximum}$$

i.e. $$P_L = \frac{2}{3} P_s \tag{17.7}$$

the 'load pressure = two-thirds of the supply pressure.

Across the valve we lose $\frac{P_s}{3}$ and across each land $\frac{P_s}{6}$

The efficiency of the valve jack unit is

$$\eta_{VJ} = \frac{P_L Q_1}{P_s Q_1} = \frac{2}{3} = 67\% \tag{17.8}$$

If this had been used with a fixed delivery pump then it would have to be capable of supplying a flow when $P_L = 0$ i.e.

$$Q_{max} = C_2 (P_s)^{\frac{1}{2}} \tag{17.9}$$

The power delivered by the pump would then be

$$P_s Q_{max} = C_2 P_s^{3/2}$$

The total system efficiency is then

$$\eta_s = \frac{P_L Q_1}{P_s Q_{max}} = \frac{2/3\ P_s C_2 \left[\dfrac{P_s}{3}\right]^{\frac{1}{2}}}{P_s \cdot C_2 P_s^{\frac{1}{2}}} = \frac{2}{3/3^{\frac{1}{2}}}$$

$$= 38\% \tag{17.10}$$

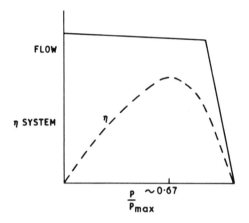

Fig. 17.2   Pump characteristics

This highlights the inadvisability of using a fixed delivery pump. In addition we have not included the pump overall efficiency. The use of a variable capacity pump with characteristics indicated in Fig. 17.2 brings the system efficiency to around the 67%. Of course using bigger valves would improve this. But can we afford the penalty? Estimates must be made to see what advantages are possible. A great deal of research work is in progress at present to develop circuits which will reduce the overall system losses. [48], [49] These could possibly involve the 'on-board' computer and self-optimizing systems [50].

17.3   Choice of pressure and components

NB All components must be compatible with fluid within the temperature bounds [13].

17.3.1   Pressure

This should be as high as possible, although often specified by external commands. As power is a product of pressure and flow an increase in pressure lowers the flow rate and reduces the size of pipes, components, velocities and losses. The disadvantages are higher leakage and eventually the fatigue factors of the components which lead to an increase in weight.

17.3.2   Servo-valves

Often selected for maximum power transmission the limit on the load is frequently set by the flow gain which is held at such a level as to maintain control. If $P_{L(max)} = 2/3\, P_s$, the flow gain is reduced to not less than

$$\left(\frac{P_s - 2/3\, P_s}{P_s}\right)^{\frac{1}{2}} = 57.7\%$$

of the 'no load' flow gain.

Accuracy, response and an adequate stability must be available together with a lack of hysterises.

17.3.3 Actuators

Here there are two basic considerations.

(1) The size should be large enough to handle the loads, that is, 'stalled' load and loads at velocity: the 'load locus'.
(2) The closed loop response of a servo is limited by the lowest open loop resonance; this is usually the load–jack natural frequency and must be within acceptable limits.

17.4 The load locus method

The characteristics of the drive should enclose the load locus [13] for an economic design, with as little overlap as possible. It is common practice to match the valve drive at the maximum power condition, the drive characteristic then being tangential to the locus as in Fig. 17.3.

The actuator is selected to match the demand for maximum force and the valve is chosen to pass the flow. The effects of increasing system pressure, actuator and valve size are shown in the diagrams. If the valve is too small or the actuator too large the efficiency is poor. If the valve is too large or the actuator too small control is difficult.

Once the valves and actuators have been selected the pumps and any accumulators can be sized.

17.5 Exercises

(A) Self-assessment

1. Even if one actuator is being used at maximum power transmission is it likely that the others are?
2. The low efficiency of the flying control units leads to a large heat generation. Can you suggest any possible methods of alleviation?
3. What variation in actuator speed would you normally expect between operation in flight and on the ground?
4. How does an accumulator aid the problem of power loss?

124

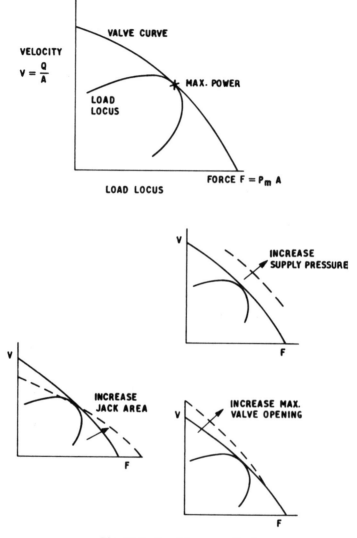

Fig. 17.3 Load locus method

(B) Numerical

1. A flying control unit comprises a fluid-powered servo-valve and hydraulic jack fed by a constant pressure supply. Show that for maximum power transmission one-third of the supply pressure is lost across the valve. Calculate the steady state positional error when a jack of 100% efficiency is controlled by a four-way valve of 'zero' lap and the desired output velocity is 3 cm/s when operating at maximum power transmission. The valve-

port opening area is proportional to the error between the desired and actual output position.

The system data is as follows:

(a) The jack has a through piston rod and can support a 'stall' load of 23 100 N at a maximum supply pressure of 210 bar.

(b) The area of the valve opening at each port is 2 mm² when the output error is 0.5 cm.

(c) The flow through a valve port is given by:

$$Q = 13 \cdot a \cdot (\Delta p)^{\frac{1}{2}} (cm^3/s)$$

where
a = the area of the opening (mm²) and
Δp = the pressure drop (bar). (1.07 mm)

2. Load conditions for an operating sequence are given below:

Select a valve, jack and pump to satisfy this set of conditions if the supply is around 120 bar. Force: N. 2000, 4000, 6000, 8000, 16000, 10000, 9000
Velocity: cm/s 4.0 || 6.0 || 7.0 || 7.5 || 8 || 4.2 || 2.5
(jack diameter ≈ 5 cm, 160 cm³/s at 40 bar; 1.92 kW).

3. A simple hydraulic system consists of a four-way valve connected to a hydraulic ram. The ram piston has an area of 39 cm². The supply pressure is 80 bar and the specific gravity of the oil is 0.875. Each valve control port has an area of 0.26 cm² per cm of spool travel. Determine

(a) the pressure drop per control port for maximum power transmission;
(b) the oil flow rate in litres/s under these conditions if the spool displacement is 0.08 cm, the discharge coefficient is 0.61; and
(c) the ram velocity and output force under these conditions.

(13.3 bar)(0.07)(1.07 m/min) (20 800 N)

18
Stability Networks, 'On–Off Servos' and Non-Linearities

(*NB* This chapter leads on to a more advanced study of the subject)

18.1 Introduction

This final chapter introduces the methods for dealing with two advanced topics, namely how, the servos can be modified to increase the damping factor without any appreciable reduction in performance, and the effects of the many non-linearities. In particular, it provides references to the literature for the design to be modified and take account of the real nature of the components and their environment. Much of this work is covered in McCloy and Martin [14] and Guillon. [51] As far as increasing the damping factor ζ is concerned this is easy to carry out on electrohydraulic units adding to or modifying the electrical networks. It is possible on mechanically connected servos to simulate the same effect, although in truth it is much more difficult. We will, therefore, merely discuss the modifications in general theoretical terms and examine how they alter the equations and block diagrams.

18.2 Network modifications [3]

18.2.1 Positional feedback

The block diagram is given in Fig. 18.1(a) and merely illustrates the case considered in detail in chapters 12, 13 and 14 [3]. In the main transfer block

$$G(s) = \frac{1}{s\left(\dfrac{s^2}{\omega_h^2} + 2\zeta\,\dfrac{s}{\omega_h} + 1\right)} \tag{18.1}$$

$$\overline{K} = \frac{1}{\tau} \tag{18.2}$$

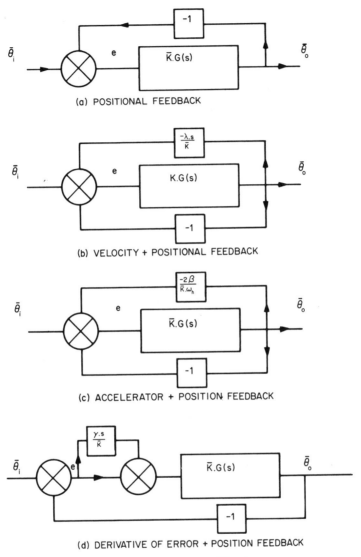

(a) POSITIONAL FEEDBACK

(b) VELOCITY + POSITIONAL FEEDBACK

(c) ACCELERATOR + POSITION FEEDBACK

(d) DERIVATIVE OF ERROR + POSITION FEEDBACK

Fig. 18.1   Network modification

For marginal stability

$$\zeta > \frac{1}{2\omega_{h\tau}}$$

(18.3)

18.2.2   Velocity feedback

Here in addition to the output position we also feed back a fraction of the output velocity ($K_1 s \bar{\theta}_0$) (Fig. 18.1(b))

where
$$K_1 = \frac{\lambda}{\overline{K}} \qquad (18.4)$$

λ being a fraction. Now

$$e = \bar{\theta}_i - \bar{\theta}_0 - \frac{\lambda s}{\overline{K}} \bar{\theta}_0 \qquad (18.5)$$

The condition for marginal stability is now

$$\delta > \frac{1}{2\omega_h \tau (1 + \lambda)} \qquad (18.6)$$

Thus the damping necessary for stable operation has been reduced by

$$\left(\frac{1}{1 + \lambda} \right).$$

The steady state error is no different to that in section 18.2.1.

18.2.3 Acceleration feedback

Here a fraction of the acceleration K_2 is fed back (Fig. 18.1(c)). In actual fact the load pressure can be used as:

$$P_L = \frac{M}{A} s^2 \bar{\theta}_0 \qquad (18.7)$$

We make

$$K_2 = \frac{2\beta}{K\omega_h} \qquad (18.8)$$

where β is a constant.

In this case for marginal stability

$$\varsigma > \frac{1}{2\omega_h \tau} - \left[\frac{\beta}{2} \right] \qquad (18.9)$$

Once again the damping necessary to produce stability is reduced.

18.2.4 Derivative of the error signal

This is shown as the extra loop within the main loop in Fig. 18.1(d). The proportion of the derivative of the error signal is

$$\left[\frac{\lambda}{\overline{K}} \right].$$

This produces a situation where for stability

$$\varsigma > \frac{1}{2\omega_h \tau} \left[\frac{1}{1 + \gamma} \right] \qquad (18.10)$$

Here again the necessary damping has been reduced by

$$\left[\frac{1}{1+\gamma}\right].$$

18.3 The 'on—off' servo

In this arrangement the valves are either fully open or completely closed (Fig. 18.2). This means that the flow rate is only dependent on the load pressure; the supply pressure is constant. It is widely used on aircraft particularly for nose-wheel steering where it is often called a 'bang-bang' servo. [52] If high inertia is involved trouble can be encountered due to the problem of stopping the movement without high pressures developing but a 'brake' or 'relief' valve can be inserted to overcome this difficulty.

A method of operation somewhat similar to this is the 'acceleration switching' servo. This uses an electrical control signal which is pulse-width modulated. The flapper is then driven hard over in alternate directions. This produces an acceleration error servo which is stabilized by adding a phase advance network. It does, however, improve the valve reliability in the presence of isolated dirt particles. [53]

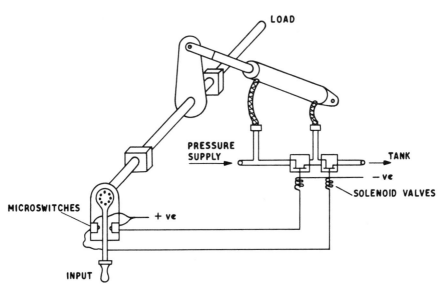

Fig. 18.2 The simplest form of 'on—off' hydraulic servo

18.4 Non-linearities — towards a solution

The most difficult problems associated with hydraulic servo design are connected with non-linear performance, that is when we cannot decide the behaviour by means of linear differential equations and transfer-functions. [13], [51] The major techniques for dealing with such behaviour are:

1. Small perturbations — linear analysis.
2. Piecewise linearization — chopping up into small linear segments.
3. Describing function analysis — determining the response to an equivalent fundamental frequency sinusoidal input.
4. Phase plane — a graphical technique (tedious).
5. Analog computer.
6. Digital computer analyses — the most modern approach using a 'step-by-step' method.

18.5   Describing function analysis

This is an approximate method described in detail in Merritt [13] and depends on assuming a sinusoidal input which produces a fundamental sinusoidal output. It requires a Fourier analysis of the output.

18.6   The results of non-linear performance

1. Fixed input — output oscillations.
2. Sinusoidal input — output jumps.
3. Sinusoidal input frequency ω — output oscillation at 2ω, 3ω etc.
4. Sinusoidal input frequency ω — output oscillation at $\omega/2$ $\omega/3$ etc.
5. Sinusoidal input frequency — output oscillations with a combination of frequencies.

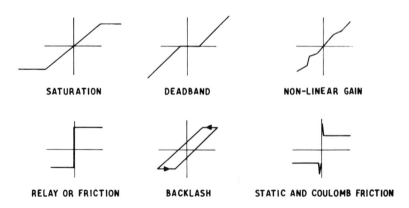

SATURATION   DEADBAND   NON-LINEAR GAIN

RELAY OR FRICTION   BACKLASH   STATIC AND COULOMB FRICTION

Fig. 18.3   Some non-linearities

18.7 Types of non-linearity

1. Saturation.
2. Deadband.
3. Non-linear gain.
4. Backlash.
5. Hysteresis.
6. Relay type.
7. Friction, for example, Coulomb.

All these produce undesirable effects and necessitate a detailed study and analysis (Fig. 18.3).

18.8 Exercises

Self-assessment

1. Can you think of the way in which acceleration feedback is carried out on a circuit?
2. Consider how nose-wheel steering can be operated using an 'on–off' servo and devise a circuit.
3. How would you recognize the presence of some non-linearities?

References

1. Conway, H. G. (1957). *Aircraft Hydraulics, Vol.1 (Systems)*, Chapman and Hall, London.
2. British Standards M24 (Aerospace) and 2917 (Industrial).
3. Stringer, J. (1980). *Hydraulic Systems Analysis;* Macmillan, Basingstoke
4. Conway, H. G. (1957). *Aircraft Hydraulics, Vol 2 (Components)*. Chapman and Hall, London.
5. Conway, H. G. (1974) *Fluid Pressure Mechanisms*, Pitman, London.
6. *Proceedings of International Conferences on Fluid Sealing (All)*, British Hydromechanics Research Association, UK.
7. Royal Aeronautical Society Data Sheet (1962) 01-01-24; 'The compressive stress of hydraulic jacks'.
8. Persons, P. and Saunders, H. (1949). The strength of jacks in compression, *Aircraft Eng.*, **April.**
9. Green, W. L. (192). Slowing down hydraulic cylinders, *Hydraulic Pneumatic Power* (UK), **November.**
10. Damasco, F. and Spon, C. (1961). High temperature fluids, National Conference on Industrial Hydraulics, Paper 15, **October.**
11. Wilson, W. E. (1950). *Positive Displacement Pumps and Motors*, Pitman, London.
12. Blackburn, J. F. *et al.* (1960). *Fluid Power Control*, Chapter 4, MIT Press, Massachusetts.
13. Merritt, H. E. (1967). *Hydraulic Control Systems*, John Wiley, Chichester.
14. McCloy, D. and Martin, H. (1973). *The Control of Fluid Power*, Longman, London.
15. Engineering Sciences Data Unit, (Losses), 69016, 66027, 66028, 67040, 72011, 72010, 69022 (UK)
16. Dransfield, P. (1967). Hydraulic lock with single land pistons, *Proc. Inst. Mech. Eng. (Lond.)*, **182,** Part 1.
17. McCloy, D. and McGuigan, R. H. (1965). Some static and dynamic characteristics of poppet valves, *Proc. Conf. Advances in Auto Control, Inst. Mech. Eng. (Lond.)*, Paper 23.
18. McCloy, D. and Beck, A. (1967). Cavitation effects in poppet valves, ASME, Paper 67FE27, Fluids, Div. Meeting, Chicago.
19. Feng, T. Y. (1959). Static and dynamic control characteristics of flapper nozzle valves, *Trans. ASME* **81** Series D.
20. Benson, R. E. G. (1967). Solenoids in fluid power, *Fluid Power Int. (UK)*, **April.**
21. Green, W. L. (1978). Flow control valves, *Hydraulic Air Eng.*, **May.**

22. Green, W. L. and Wood, G. (1973). Some causes of chatter in direct acting spring loaded poppet valves; BHRA 3rd Fluid Power Symp., Turin.
23. Green, W. L. (1972). The stability of poppet relief valves, *Hydraulic Pneumatic Power,* **September.**
24. Green, W. L. (1968). The poppet relief valve—a scientific approach to design, *Hydraulics and Pneumatics (USA),* **December.**
25. Ma, C. Y. (1966). The analysis and design of pressure reducing valves, *Trans. ASME,* Paper 66, WA/AD-4.
26. Green, W. L. (1973). Twenty years of aircraft hydraulics, *Hydraulic Pneumatic Power,* **June.**
27. Fauvre, M. Le (1970). Concorde hydraulic system, *Fluid Power Int. Conf.,* **June.**
28. Anon (1970). Harrier V. STOL., *Fluid Power Int.,* **September.**
29. Aerospace 1983, Fluid Power and Control Systems SP-5546 Aerospace Congress, California Oct 3–6 (Auspices Soc. Automotive Engineers USA)
30. Green, W. L. (1973). The effect of discharge times on the selection of gas charged hydraulic accumulators; BHRA 3rd Fluid Power Symp. Turin.
31. Green, W. L. (1979). Accumulator time constants and the index method, Int. Conf. on Fluid Power, Chicago.
32. Shute, V. A. and Turnbull, D. E. (1961). Some trends in electro-hydraulic servo valve design, Proc. Oil Hydraulics Conf. I.Mech.E. London, Paper 11.
33. Martin, H. R. (1967). A non-mathematical look at servo mechanisms, *Hydraulic Pneumatic Power.,* **August.**
34. Green, W. L. (1970), A jack servo using an open poppet valve, *Hydaulic Pneumatic Power,* **September.**
35. Hayward, A. T. J. (1967). Compressibility equation for liquids—a compressibility study, NEL. Report 295 (UK)
36. Hayward, A. T. J. (1961). Aeration in hydraulic systems, Oil Hydraulic Conf. Inst. Mech. Eng., London, **November.**
37. Magiorem, V. G. (1967). How Hydraulic Fluids Generate Air, *Hydraulics and Pneumatics.* **October.**
38. Green, W. L. and Wood, G. (1973). The analysis of resonant conditions in high pressure fluid power circuits containing a fixed delivery pump, Int. Conf. on Vibration Problems in Industry. UKAEA, Keswick, **April.**
39. Private Publication (1980). *Quieter Fluid Power Handbook,* BHRA Fluid Engineering (UK)
40. Conway, H. G. (1957). *Aircraft Undercarriages,* Chapman and Hall, London.
41. Harpur, N. F. (1953). Some design considerations of hydraulic servos of the jack type; Conf. on Hydraulic Servo-mechanisms. Proc. Inst. Mech. Eng., (UK) p. 41
42. Shearer, J. L. (1957). Non-linear analogue study of a high pressure servo-mechanism. *Trans. ASME,* **79,** 465.
43. McCloy, D. and Beck, A. (1969) Flow hysteresis in spool valves; 1st BHRA Fluid Power Symposium, Cranfield, (UK). **January.**
44. Green, W.L. (1967). Flow reaction forces, *Hydraulic Pneumatic Power,* **August.**
45. Martin, K. F. (1970). Stability and step response of a hydraulic servo with special reference to unsymmetrical oil volume conditions, *J. Mech. Eng. Sci.* **12** (5), 331.
46. Walters, R. (1967). *Hydraulic and Electro-Hydraulic Servos,* Iliffe, London.
47. Morse, A. C. (1963). *Electro-Hydraulic Servos,* McGraw-Hill, London.
48. Green, W. L. (1979). Minimising power losses in electro-hydraulic systems, *Hydraulic Air Eng.,* **January.**
49. Green, W. L., Sanger, J. and Suresh, B. N. (1980). The reduction of fluid power losses in electro hydraulic systems, *Israel J. Tech.,* **18,** 293–303.
50. Green, W. L., Sanger, J. and Suresh, B. N. (1981). On-line control for the regulation of the power source in multiple servo electro-hydraulic systems, ASME Winter Meeting, Dynamics and Control Division, Washington, Nov-81-WA/DSC-3.

134

51. Guillon, M. (1961). Hydraulic servo systems, Butterworths, London.
52. Hancock, K. G. and Persons P. (1952). Power steering for aircraft, *J. Roy. Aeronaut. Soc.* **July**.
53. Murtaugh, S. A. (1959). An introduction to the time-modulated acceleration switching electrohydraulic servomechanism. *ASME J. Basic Eng.,* **June**, 263–271.
54. De Russo, P. M., Ray, R. J. and Close, C. M. (1965). *State Variables for Engineers,* John Wiley, Chichester.

Index